"With deep intelligence and an acute and off-center sensibility, Bob Ostertag gives us a riveting and highly personalized view of globalization, from the soaring skyscapes of Shanghai to the darkened alleys of Yogyakarta."
—Frances Fox Piven, coauthor of *Regulating the Poor* and *Poor People's Movements*

"If you want an insightful, witty panorama of this brave new world we are making, follow Ostertag around it for a year—or read this book."
—Jeremy Brecher, author of *Strike!* and *Common Preservation*

"This is global reporting at its best—a pointillistic portrait of the troubles of our time rendered in riotous detail. It reads like a riff, not a sermon."
—Richard Manning, author of *If It Sounds Good, It Is Good*

"*Facebooking the Anthropocene in Raja Ampat* takes us through the grim dystopic world that we live in, but somehow we also witness the resilience of the human spirit, including its queer dimension. A joyful read."
—Dédé Oetomo, Indonesian gay activist and scholar

"A horrifying, exhilarating, hilarious and jaw-dropping glimpse of our present and future—as Ostertag details how even the most remote desert islands on our planet are becoming rapidly infected and influenced by the consumerism, technology, and plastic garbage that the rest of us have long ago surrendered to."
—George Brant, author of *Grounded*, *Elephant's Graveyard*, and other plays

KAIROS

In ancient Greek philosophy, *kairos* signifies the right time or the "moment of transition." We believe that we live in such a transitional period. The most important task of social science in time of transformation is to transform itself into a force of liberation. Kairos, an editorial imprint of the Anthropology and Social Change department housed in the California Institute of Integral Studies, publishes groundbreaking works in critical social sciences, including anthropology, sociology, geography, theory of education, political ecology, political theory, and history.

Series editor: Andrej Grubačić

Recent and featured Kairos books:

Asylum for Sale: Profit and Protest in the Migration Industry edited by Siobhán McGuirk and Adrienne Pine

Building Free Life: Dialogues with Öcalan edited by International Initiative

The Sociology of Freedom: Manifesto of the Democratic Civilization, Volume III by Abdullah Öcalan

In, Against, and Beyond Capitalism: The San Francisco Lectures by John Holloway

Anthropocene or Capitalocene? Nature, History, and the Crisis of Capitalism edited by Jason W. Moore

We Are the Crisis of Capital: A John Holloway Reader by John Holloway

Re-enchanting the World: Feminism and the Politics of the Commons by Silvia Federici

Autonomy Is in Our Hearts: Zapatista Autonomous Government through the Lens of the Tsotsil Language by Dylan Eldredge Fitzwater

The Battle for the Mountain of the Kurds: Self-Determination and Ethnic Cleansing in the Afrin Region of Rojava by Thomas Schmidinger

For more information visit www.pmpress.org/blog/kairos/

Facebooking the Anthropocene in Raja Ampat

Technics and Civilization in the 21st Century

Bob Ostertag

KAIROS

PM

Facebooking the Anthropocene in Raja Ampat: Technics and Civilization in the 21st Century
Bob Ostertag
© 2021 PM Press.

ISBN: 978-1-62963-830-0
Library of Congress Control Number: 2020934737

Cover by John Yates / www.stealworks.com
Interior design by briandesign

10 9 8 7 6 5 4 3 2 1

PM Press
PO Box 23912
Oakland, CA 94623
www.pmpress.org

Printed in the USA.

Hope cannot be said to exist, nor can it be said not to exist. It is just like paths across the earth. For actually the earth had no paths to begin with, but when many pass the same way, a path is made.

—Lu Xun, 1881–1936

Hope is only man's mistrust of the clear foresight of his mind. Hope suggests that any conclusion unfavorable to us must be an error of the mind.

—Paul Valéry, 1919

Contents

Preface

"Are Two Dimensions Enough?" was written fourteen years ago and considers the impact of the networked screen on the human imagination. "Facebooking the Anthropocene in Raja Ampat" was written four years ago and is a journal of thoughts written while traveling the world. I completed "Technics Turntables and Civilization" last week, tracing the trajectory of electronic music from the time I showed up playing synthesizers in the 1970s down to the present day, and the metamorphosis of the meaning of live music.

As diverse as these essays may seem, each addresses the nexus of human beings and technology. And, in each case, the method I follow is to simply observe, and listen to, real people living today. I hope I have avoided both speculation about the future and generalizations so broad that they might be bent into a causal edifice of "theory."

Technology is spreading across the world far faster and farther than ever. There are few if any people left outside the sphere of digitally networked screens. The last people to remember life before computers are alive today. I think it is worth our while to listen closely to what they have to say, before they disappear.

Just as there is nowhere outside the digital sphere, there is nowhere beyond the range of the detritus of the industrial system that maintains it. The world we are building is overtaking the world we found. This process is so far along that in some cases the only way to maintain the remnants of the world we found is to us to rebuild them synthetically. (For example, in the first essay you will meet 3D printed synthetic coral.)

We ourselves are also changing, though this is harder for us to see. The speed of technological change is impossible to miss. We cannot avoid it. Many of us now live so completely encased in the world we have built that we don't even notice the changes in the world we found, until they erupt into catastrophes like forest fires and heat waves. But it is easy to miss the changes in ourselves. We imagine ourselves to be holding still while the changing world rushes past.

I am often told that my frankness and honesty about all of this veers into "hopelessness" or "pessimism." But I find no hope in dishonesty. We cannot lie to ourselves and not know that we are doing so. I am more interested in love, inspiration, and humor. And I find them plentiful. Even in these dark and untethered times. Because of these dark and untethered times.

Bob Ostertag
January 2020

Facebooking the Anthropocene in Raja Ampat

Notes on a One-Year Concert Tour

Dear Friends,

Forgive me for being such a poor correspondent during my year around the world. The daily kaleidoscope was so relentlessly compelling I found it difficult to collect my thoughts, much less write them down.

At the end of it all I have arrived in Berlin, an island of sanity in a crazy world.

More precisely, I am in the naked gay sunbathing section of the Tiergarten, Berlin's large city park. Green grass, green trees, blue sky, a lovely blanket of calm everywhere, lots of naked men lying in the sun. Families with little kids walking by. The easy mixing of naked gay men and kids on their way to get ice cream bothers no one.

Beautiful Berlin.

The only men not lying motionless are me and two others just to my left. One of these is in a motorized wheelchair. He has a disorder that makes it impossible for him to control his movement and is flailing around in the chair. The other man is the flailing man's caretaker, who is spoon-feeding the one in the chair. The man in the chair is naked. His caretaker, clothed, has brought him here so he can be naked in the park with the rest of the guys.

The condition of the man in the chair is so severe that it is hard for him to swallow, and the quiet tranquility of the park is periodically broken by his loud and urgent choking when the food goes does the wrong way.

I am doing yoga.

Perfect.

At last I can think.

Love,

Bob

The first thing that jumps out at you while traveling the world today is that the world you are moving through is coming apart. Or rather, there is one world—the old world—coming apart, and a new world is coming into view.

I have seen this with my own eyes, laid out in front of me like a buffet, yet I must almost certainly be wrong. The end of the world has, after all, been described over and over through the millennia by honest and well-meaning people who were certain they were witnessing it and who convinced many others to believe the same.

William Miller, for example, did a particularly deep dive into Bible numerology, and concluded that the world would end "sometime between March 21, 1843, and March 21, 1844." The message spread first through a periodical out of Boston called *Signs of the Times*, then by a Millerite movement. By the time the moment for the end of the world arrived, the movement claimed to have distributed five million copies of Millerite periodicals, and there were avowed Millerites in Canada, England, Norway, Chile and the Sandwich Islands (now Hawaii). When March 21, 1844 passed uneventfully, Millerite numerologists redid their math, found the mistake (it was in the 2,300-day prophecy in Daniel 8:14) and *proved* that the world would end on "the tenth day of the seventh month of the present year, 1844," which worked out to be October 22. When the day came and went, what followed became known as "The Great Disappointment." Stunned Millerites confronted not just their own confusion, but church burnings, mob violence, and constant ridicule.

So no one should be dumb enough these days to declare that the world is ending. Yet, even as the naked man in the wheelchair was choking on his food, the world's preeminent geologists were, in fact, announcing the end of the old world and the beginning of the new, in the form of a new geological epoch, the Anthropocene: a world in which the consequences of human activity have reached geological proportions.

More precisely, the Anthropocene Working Group, of the Subcommission on Quaternary Stratigraphy, voted to recommend that the International Commission on Stratigraphy declare the Anthropocene a new geological epoch.

Human activity, they declared, now affects natural global systems and processes so profoundly as to leave its mark in the stratigraphic record. Martian stratigraphers arriving on earth millions of years in the future to study the chemical composition of layers of rock would learn that vast and unprecedented planetary changes suddenly began occurring right about now.

Geologists spend a lot of effort sorting out when one geological epoch ends and a new one begins. They have agreed on dividing lines for geological eras going back 545 million years: the Pleistocene, the Pliocene, the Miocene, and so on. Each of these geological epochs lasted for millions of years. The most recent epoch, the Holocene, began just eleven thousand years ago. By the timescale of all other geological epochs, the Holocene was a newborn. So to declare that the Holocene has already given way to the Anthropocene is to declare that something has occurred (a geological era has lasted just ten thousand years) that has no precedent on earth in the last five hundred million years: Earth 2.0.

Geologists are a pretty staid and stable bunch. They arrived at this utterly preposterous conclusion by analyzing the chemical composition of things like glacial ice cores, layers of rocks, and sediment at the bottom of lakes. But nongeologists without chemistry labs can still see the processes of the Anthropocene at work. All we have to do is hop on an airplane and go look around.

Incidentally, airplane travel is one of the most effective ways an individual can contribute to the processes of the Anthropocene. The only behavior a single person can engage in that leaves a bigger carbon footprint than airplane travel is eating beef.

I left my home in San Francisco in March 2015. California was in the midst of a historic drought. Lakes were drying up and forests were burning. The biggest forest fires in history. Just before I left, the crab fishing season along the entire coastline of California was cancelled because the unusually warm waters in the Pacific had turned the crab toxic. As I was coming home a year later, research confirmed that the largest coral "bleaching event" ever was underway in that same Pacific Ocean.[1]

I arrived in Europe for the historic European heat wave of 2015. Temperatures topped 100 degrees Fahrenheit in northern Germany. I left Europe for Lebanon, where one in five people living within the nation's borders is a refugee from neighboring countries racked by historic droughts.

In El Salvador I attended a water management class for campesinos from villages with no more water. Their water now arrives by truck. Central America was in its worst drought on record. The corn harvest failed for many Salvadoran subsistence farmers for the first time in anyone's life. The purpose of the class was to search for new water management methods that would allow subsistence farming to continue in areas where the only water available arrives by truck.

How can you have subsistence farming if the only water arrives by truck?

I visited my friend Myrna at her beach house outside of San Salvador, the capital city. The house formerly opened onto a beautiful sandy beach, but a few months before my arrival the area experienced what was being called a "tidal event." One day the high tide rose higher than it ever had or was predicted to, leaving both the beach houses of the comparatively well off, and the crude wooden shops of their poorer neighbors, standing in saltwater. Not nearly as dramatic as a tsunami. The water arrived on a tide instead of a wave. Nevertheless, the "event" covered the land with salt water and the beach with rocks. The salt killed the trees, while the rock now completely blankets the beach. Not the tiny sort of rocks that generally cover rocky beaches but boulders of all shapes and sizes. Boulders strewn about everywhere, one on top of another on top of another.

For the army colonel who owns the only big modern hotel on the beach and has strings to pull and money to spend, there is no problem. If he can marshal the resources to move the rocks off his property, he will have the only beach around. Otherwise, he can take his money and his connections elsewhere. For middle-class Myrna, the rocks mean that the beautiful beach she came here for is gone. Not a happy turn of events, but she will survive. It is her beach house, not her home. But for all the shopkeepers who had invested what little they had in the ramshackle wooden shops that lined the beach, it is the end of their livelihoods, and they live in a very poor country where alternative livelihoods are hard to come by.

On the scale of the consequences of the transition from the Holocene to the Anthropocene, the "tidal event" on El Salvador's coast is no big deal. Tinier than tiny. Not even anything. Yet there, where the beach used to be, sit tons and tons of rock and the abandoned shacks left behind by their proprietors, along with all their hopes and dreams.

What will the displaced shopkeepers from the beach do? Or the campesinos from the mountain villages which have run dry? The best option may be to flee, and begin the long and dangerous journey to the US, where they will be viewed as "illegal immigrants," not "climate refugees."

Prediction: in the near future, a new political drama will emerge over who gets to be a "climate refugee." Until now, the big political drama of human migration has been about who counts as a "political refugee" fleeing political oppression, as opposed to those fleeing economic hardship, who get stuck being "immigrants." As if economic hardship and political oppression could be so neatly disentangled. Was the economic hardship the cause or the result of the political repression? International NGOs, United Nations officials, and academics have actually been trying to develop criteria to answer that question. But in practice the distinction is made not by reasoned argument but by political power and advantage. In the US, people who flee regimes that are not friendly with the US are "refugees," while those fleeing friendly regimes are "immigrants." "Refugees" are sometimes welcomed because they demonstrate the superiority of our political system. "Immigrants" are tolerated up to a point as long as they clean our toilets, pick our crops, care for the children we are too busy to care for ourselves, and don't bother anybody. With Donald Trump in the White House, the status of immigrants is being profoundly contested. Though neither Trump nor his supporters have explained who will clean their toilets or their children if all the undocumented immigrants are deported.

A third character will soon be added to this drama: the "climate refugee." There will be a *lot* of potential candidates, for climate change will create migration on a scale never known. But how will we distinguish the cause of their hardship in a world where poverty and war were endemic even before climate change came along?

The exodus has already begun. Nearly all of those risking everything to get into Europe today are from drought-stricken areas. Those areas from which they fled are, in turn, full of even more desperate people who fled even more severe droughts. Get ready for protests and demonstrations, slogans on banners, angry denunciations, and streets full of riot police and tear gas—all about who gets to be declared a "climate refugee." Coming soon to a screen near you. Hint: subsistence farmers from drought-afflicted areas in Central America and North Africa are not likely to make the cut; oil billionaires from the Persian Gulf, an area which is predicted to be too hot for human life by the end of the century, are a better bet.[2]

I went from El Salvador to Detroit and landed in yet another water crisis. But Detroit was an outlier among the world's water crises, because its water crisis was entirely the creation of policy. There was no drought or tsunami or even tidal event. To the contrary, Detroit lies at the edge of the Great Lakes system, which contains 21 percent of the world's surface freshwater, suggesting this might be a good place to seek shelter from climate catastrophe. Yet many of Detroit's poorest residents were without water due solely to government policy.

At the peak of American industrial power in 1960, Detroit had a population of 1.7 million. When I arrived less than half that number remained. The jobs had migrated to southern states where unions were not a problem, or to other countries. Many whites had migrated to the burbs. The process had been going on for so long that much of the debris from the collapse had been carted away. You drive through entire neighborhoods of empty blocks with absolutely nothing on them, the flat and barren landscape more reminiscent of an Iowa cornfield in winter than a city. You could farm that land if you wanted to, and some committed urban farming collectives were doing so. You just couldn't plant anything in the dirt, which is contaminated with lead. You have to truck in clean dirt.

Detroit declared bankruptcy in 2013, the largest American city in history to do so. An "emergency financial manager" was appointed by the Republican governor of Michigan and tasked with somehow squeezing enough blood from the stone that is Detroit to pay off the city's creditors. He decided one way to go about this was to collect outstanding water bills, by shutting off the water to those who owed more than $150.

Figuring out who in Detroit owes more than $150 on their water bill is not as simple as it might sound. The Detroit water system was built for a population heading toward two million but serves less than one million. The fixed costs of the giant system are now divided among less than one million users, making for water bills nearly double the national average in one of the poorest cities in the country. And then there are all the leaks and breakages from the wreckage of the old system. One person was presented a water bill of $10,000.

City-owned golf courses, however, owed more than $400,000 but were under no threat. Same for the professional hockey and football arenas ($137,000). Collecting just these three bills would have raised as much money as shutting off the water to 3,500 people.

You can live in the dark, and you can live cold. You can live hungry, but you cannot live without water. You cannot even go a full day without water.

If there is no water in your house, the state can take your children away. Thousands of people were put at just that risk, to raise the same money that could have been raised collecting three bills from sports facilities.

The United Nations sent a special representative to Detroit to inform the city leaders that access to water is a basic human right, and that the city's water policy violated international human rights law. The dumbfounded authorities replied that this was the United States, and no one in the world tells Americans what to do, and weren't UN representatives supposed to be in some other country where there were real human rights issues?

The problem goes beyond water. All the utilities are a mess. Over the years during which over a million people left, properties were abandoned, occupied, stripped for parts, and occupied again. In many areas only the poorest of the poor, the ones who simply had nowhere to go, remained. They lived where they could. Just figuring out who owns a house can be vexing. There are many people paying rent to someone who claims to be their landlord but doesn't actually own the house. Then one day the "real" landlord shows up demanding back rent.

Winters in Detroit can be brutally cold. The icy wind sweeps in off Lake Erie, and snow piles up along the roadside higher than the cars. In April I met a woman who had just lived through the long, dark winter with all her children in a house for which she paid rent but that had no electricity or heat. They huddled in the freezing dark. After surviving the ordeal, one fine spring morning she received a utility bill for $600, for gas and electricity she had not used. I met her at the office of the Michigan Welfare Rights Organization, in utter despair.

I was in that office accompanying Maureen Taylor through just one of her grim days running the Michigan Welfare Rights Organization. Born and bred in Detroit, built like a fire hydrant, she is that rare gem of a person who can work against overwhelming odds, fighting the most outrageous oppression, in situations of near suffocating despair, for decades on end, and still wake up the next morning busting for a fight. I am in awe of her.

She told me about teaching social work at Detroit's community college.

> I told my students there was one class they could not miss. If they missed that day, they failed. It was the day I took them to the morgue to watch an autopsy of an unidentified corpse.
>
> They would say, "Oh, Miss Taylor, I can't handle this. I am going to throw up."
>
> And I'd say, "That's all right, you just throw up right here."
>
> "No, Miss Taylor, I am going to faint."

"That's OK, you just faint right here."

Because you see, if you want to do social work in Detroit you have to know what death is.

In China I breathed some of the world's worst air. Just before my concert Beijing issued an air pollution "red alert" for the first time ever. Thousands of schools closed. Thousands of factories shut. Everyone on the street wearing face masks. But you didn't go outside at all if possible. This was official policy: do not go outside if you can avoid it. The capital's economy screeched to a halt, which is not permissible in a globalized world. Especially if the capital in question is in China. The threat of more red alerts in the future put an urgent question mark over whether the factories at the heart of the globalized world could reliably function, or whether workers could even get to work. The government addressed this urgent concern by raising the threshold for how bad the air has to be for a red alert to be issued.

Incredibly, red alert days are not that much worse than every other day in Beijing. The taste of smoke in your mouth is the last thing you are aware of when falling asleep and the first thing you are aware of waking up. Your pillow and sheets and clothes all smell of smoke. You wake up and wonder if something is burning. The answer is yes.

Go to the park in Beijing and you can watch groups of elderly retirees passing the time in directed group deep-breathing exercises while busy middle-aged people in corporate attire hurry past on their way to the next meeting, face masks covering their nose and mouths. The silly old geezers can't keep up with the times.

Beijing isn't even among China's ten most polluted cities. There are cities where millions of people live and the air is even worse. I did not go to any of those cities. I am not a timid traveler. I have wandered through areas of drought, civil war, drug gangs, corrupt dictatorship, and more. But I don't want to go to those Chinese cities. I don't want to breathe the air.

In Vietnam, Hanoi's air is right behind Beijing's. Visibility is even worse. Looking out from the luxury restaurant at the top of the city's one new skyscraper (the Communist capital's tenuous claim to having arrived in the New Asian Century), visibility was about one kilometer though it should have been a cloudless, sunny day. Beyond that everything disappeared into a toxic cloud. Just a few years ago Hanoi was compared with Havana as a charming backwater that had escaped the worst ravages of neoliberal development due to the economic failures of communism. China might have stunned the world with the speed with which it went from "underdevelopment" to third-millennium industrial nightmare, but Vietnam is doing it even faster. That's right, the movement whose heroic battle against US imperialism at its zenith inspired the global anti-imperialism of the 1960s has now given the world an apocalyptic morass of environmental catastrophes created in record time. As I write these words, an unprecedented fish die-off has blanketed 120 miles of Vietnamese coast with toxic fish carcasses.[3]

Traffic in many Asian cities has reached a breaking point. Traffic has long been bad, but the scale of today's traffic catastrophe is new. In Beijing, local friends who thought they knew their own city chaperoned me around. But every time they would calculate the time it would take to get to where we needed to go, they would be off by an hour or more. I arrived at my own lecture at the Music Conservatory two hours late. Lifelong residents miscalculating traffic by hours every day— that is how quickly the traffic situation had deteriorated.

The traffic in Jakarta, Kuala Lumpur, and Bangkok are as bad as Beijing. Manila is worse. I don't know how to describe Manila's traffic. I honestly don't understand how the economy doesn't just come to a stop. Perhaps it will.

I was in Manila during an economic summit attended by numerous heads of state. The street closures which made it possible for the likes of Barack Obama and Vladimir Putin to move through the city made movement even more difficult for everyone else. I needed to run an errand to see a friend six kilometers from my hotel. Each day my hosts told me, "No, do not try to go today, the traffic is too bad. Wait until tomorrow." When my days in Manila ran out I simply had to attempt the errand. It took four hours in a cab to travel six kilometers. I could walk faster, but there is no walking in Manila. There is no sidewalk. Cars wall to wall. If you walked, you would die.

In fact, there is no walking in many Asian cities. In China I met activists from a group called "Walk in the City," who are trying to reintroduce walking to China's urban population. In Surabaya, on the eastern tip of Java, I stayed with a "walking activist" who was saving money to fly to the first pan-Asia "walking conference" in Hong Kong.

Even the weather was wrong. Throughout Southeast Asia rain was either falling in enormous amounts when it typically did not, or not falling when it usually did. Sydney got a typical month's worth of rain in ninety minutes. And so on.

Underwater, things are even worse than up above, if that is possible. In the Philippines I went to a remote island my Filipino friends recommended as "pristine," a beautiful seascape as far removed from the toxic mess that is Manila as can be imagined. And it was indeed stunningly beautiful. Crystal clear water. Sculpted limestone islands whose incredible shapes rival the most fantastic icebergs. Deserted beaches. The first day I went on a guided snorkeling trip to a reef made famous by Jacques Cousteau. All the tourist promotional materials feature the same Jacques Cousteau quotation proclaiming this place one of the top snorkeling spots in the world. Snorkeling around the reef gives you some idea of what he was excited about. It is extensive and diverse, the shallow depth makes it easily accessible to a snorkeler, and to the sunlight that penetrates the shallow water, making the colors especially brilliant.

The reef is also almost entirely dead. A vast expanse of whitewashed coral bones. A coral graveyard. Almost no fish. At one end of the reef there is a bit that is still alive. It is stunning. Deep purples and siren reds and verdant greens. Brilliant orange and elusive silver. But beyond this tiny refuge, the reef is nothing but old bones.

When I exited the water, my ever-attentive guide was eager to hear my thoughts.

"You took me to a dead reef."

His expression was so pained I regretted saying the obvious. Of course he knew that the reef was dead.

Until recently he had been a single-line fisherman, which is a life of poverty. You sit in your little boat with your fishing pole in hand as the highly capitalized Japanese boats race past and suck up all the fish with these sort of hi-tech vacuum cleaners of the sea. Or the pirate fishermen throw bombs. Yes, bomb fishing is still practiced around here. Throw a bomb in the water and see what dead things float to the surface. Kills everything, including the coral. I heard them going off in the distance.

Tourism arrived in this place over the last five years and changed my guide's life. He and his family live better working for tourists than they ever could from fishing, though he doesn't really understand tourists. Flying around the world to look at coral reefs isn't something his people do. But he is proud that his humble little corner of the world now draws people of unimaginable wealth from the farthest reaches of the globe to marvel at the natural beauty, and his livelihood now depends on these visitors returning home to tell their friends they had a fantastic time, or more importantly leaving a positive review on a crowd-sourced travel website.

Guides take tourists to this reef because they can advertise the trip with the words of some dead guy named Jacques Cousteau from the land where the rich tourists come from. So taking tourists to that reef is what he is going to do, and he is going to do the best job of it he can. He is going to be sunny and upbeat and talk of all the natural beauty of his home. Hopefully, not too many clients will point out that the reef is dead. Some of those in my group knew so little about coral that they didn't even notice.

His only alternative at this point is to go back to single-line fishing, but the fish are disappearing. And what about the Japanese vacuum cleaners and the fish bombing?

And anyway he is a Seventh Day Adventist, which means he believes "God does not want me to eat shellfish." The King James version of the Bible put it like this: "Whatsoever hath no fins nor scales in the waters, that shall be an abomination unto you." In the same passage God instructs that animals with cloven hoofs and cuds they chew are permissible, but those with just cloven hoofs or cuds they chew are verboten. It's got to be both or neither, not one or the other, which leaves out camels, rabbits, hyraxes, and pigs. Locusts, beetles, and grass-hoppers are OK, "but all *other* flying creeping things, which have four feet, *shall be* an abomination unto you." The dietary abominations of a desert religion two thousand years ago, packaged and transported by missionaries and colonial armies, washed up on shore of a Southeast Asian beach paradise.

Seventh Day Adventists, by the way, are what became of the Millerites after the world failed to end according to forecast. Many at the time thought that would be the end of it, but no. They sorted out that Miller had been correct about the date for the Second Coming, just wrong about where Jesus was coming to. That was the day Jesus came to cleanse heaven, not earth.

On the day my guide was explaining to me why God did not want him to eat the shellfish all around him, the Seventh

Day Adventist Church had 19 million baptized members in 215 countries, making it the twelfth-largest religious body in the world, and the sixth-largest highly international religious body. The church is considered the iconic case of "True Believer Syndrome"—how people continue to believe in a paranormal event even after it has been disproven.

★

My last stop in Southeast Asia was Raja Ampat. If anywhere in the world can still be considered remote, Raja Ampat is remote. To get there, you first must travel to a major Indonesian city of your choice. From there you get a flight on a small plane to Sorong, on the island of Papua. From there you take a two hour ferry ride to the town of Waisai. And from there you go by motorboat for another hour or two and you are in the Raja Ampat archipelago.

Raja Ampat is known for having the last relatively healthy marine ecosystem in the world, the result of its remote location, ocean currents, and weather. People come from all over the world to scuba dive here, because, as one diver said to me, "Who wants to dive in the Caribbean anymore?" Thailand is overrun. Florida was ruined decades ago, and so on. Which leaves Raja Ampat as the world's undisputed premiere dive destination. You can strap on an air tank and spend a few glorious days imagining that the oceans' ecosystems are not collapsing.

The remote location makes it difficult to run a dive resort at the level of luxury those who travel the world to dive apparently expect. Amenities like beer, food (other than fish), and fuel must be brought in from far away. So my week diving in Raja Ampat, one of the poorest regions in the world, was one of the most expensive weeks of my life.

The underwater world I found there is indeed spectacular. I dove through shimmering swarms of fish whose movements combined into dazzling emergent shapes and patterns. Swam with sharks and eels and octopus and so much more. Marveled at the shapes and colors of coral. Looked eyeball to eyeball with a giant manta ray of about six thousand pounds and twenty feet across.

I also traveled through Raja Ampat above water by kayak for ten days with a Papuan guide named Sedik, paddling from village to village and sleeping in "homestays"—bamboo shacks built to accommodate the few tourists who have just started to show up. Here I was really out on the edge of tourism. Paddling a fiberglass kayak, a hi-tech descendant of the traditional sea going vessel of the Arctic, but on the equator alongside locals paddling dug-out canoes. One night there was no village within reach, and we camped on the beach of an uninhabited island, a "deserted tropical island" fantasy come to life. The quiet. The tropical breeze. Palm trees. Sculpted limestone rock. Plentiful fish. A deserted beach. And garbage.

Lots and lots of garbage. A staggering, mind-numbing amount of garbage. Old flip flops and medicine bottles and soap bottles and shampoo bottles and soft drink bottles and motor oil bottles and aerosol bottles and specimen cups and so on. The beach had a carpet of garbage. There was literally no place to put your foot down without stepping on garbage. It covered everything.

When I think of the Anthropocene, I most often think of rising carbon dioxide levels and loss of biodiversity, but geologists argue that plastic is just as important. They write papers with titles like "The Geological Cycle of Plastics and Their Use as a Stratigraphic Indicator of the Anthropocene."[4] Little bits of plastic now cover the earth; from mountain tops to sea floor. Over the next few thousand years it will all get compressed into a geological layer which will then get buried over the next few million years. Those Martian stratigraphers arriving millions of years from now will know from studying rock layers that something previously unknown on earth suddenly blanketed the planet at this time. We call it plastic.

We landed our kayaks on the trash, and I watched in confusion as Sedik went to work gathering up all the downed palm branches that lay about the beach. He stacked them into one large pile, and then set it on fire.

"What are you doing?" I asked.

"Cleaning the beach."

I was stunned.

It was true that when we landed our boats there were dead palm branches all around that no one would describe as beautiful. You quickly realize that the deserted tropical beach that Westerners dream about—and gaze at longingly on their virtual desktops and in beer commercials—is not deserted at all. It is a beach where someone regularly cleans up the downed palm branches. I got that. But we were standing ankle deep in a carpet of plastic garbage. The palm branch fire itself was burning on a bed of plastic garbage. But the garbage did not bother Sedik. He seemed to not even see it.

I got to work doing my own beach cleaning, clearing plastic from a small section of sand. Sedik watched with curiosity. What in the world was this strange visitor doing? I soon discovered that clearing the plastic from even a tiny little patch of beach was beyond the ability of one person. It was like doing amateur microscale stratigraphy. Intact artifacts like plastic bottles formed the top layer. Clearing that away revealed a layer of fragments of the same artifacts that were intact in the top layer. Excavating layer by layer, the fragments became progressively smaller until they became indistinguishable from the sand. Maybe the sand itself was plastic. Below that would be the plastic fragments even smaller than the sand. Microplastic, it is called. I couldn't see it because I didn't have a microscope, but it was there.

And anyway, even if I was able to collect all the plastic from my little patch of beach, what would I do with it? It wouldn't fit into my kayak. No one was going to take a motorboat all the way out here to collect the garbage. And if they did, where would they take it? To another island? I decided to try burning it, which turned out to be a really stupid idea. After the truly horrendous black smoke had cleared, where bottles and cups and so on had been, there was now a gnarled mass of icky black goo.

I thought back to the previous night on the nearest inhabited island, where I slept in a bamboo hut decorated with lines of string hanging from an awning over the porch. Knotted in the strings every few inches were beautiful seashells, alternating with colorful bits of plastic garbage and even the occasional rusty aerosol can. The proprietor seemed to make no distinction between seashells and garbage. To him it was all interesting and colorful stuff collected from the environment. Gifts from the sea.

I watched Sedik, squatting on a carpet of plastic in the sunset, machete in hand, preparing a log which he would drive into the plastic beach to support our hammocks for the night. I looked at the forlorn results of my own labors: the little patch of exposed grains of sand and plastic, and the black goo left behind by my inept fire. Of the two of us, Sedik was the "modern" one, the one who was ready for life in the Anthropocene. The "primitive" one was me, clinging to a past that is long gone.

While contemplating all of this after I returned home, I read up on microplastic. Turns out microplastic comes in two varieties. "Secondary microplastics" are larger pieces of plastic that have broken down to microscopic size over time, like the layers I had unearthed on the beach. "Primary microplastics" are particles of plastics that are *designed* to be microscopic. Primary microplastics are used in cosmetics, like sunscreen. Dow Chemical, for example, makes plastic "sunspheres" just 0.0003 mm in size. One small tube of sunscreen might contain 100 trillion particles.

While traveling through Raja Ampat I had been vaguely aware that sunscreen was viewed with increasing concern as another industrial pollutant, so I wore long-sleeved synthetic "travel shirts" and long-legged "travel pants" to cover my skin and cut down on the sunscreen. I was not aware that these high-tech travel clothes, which are so lightweight yet provide such effective UV protection, *also* contain microplastics and shed a trail of the stuff in their wake. Both the lotion and the clothing I was wearing to protect myself from the intense sun radiation of the Anthropocene shed microplastic all over the place.

Perhaps the time has passed when white people who need protection from the sun should travel to such places.

Microplastic is now found in the vast majority of fish, in sea salt, on mountain tops, and would probably be found in me were someone to look for it. Just by bringing my own damn self to this beach I was spreading plastic.

The Great Greater Greatest Migration
Martian astronomers observing earth today with a powerful enough telescope would see that the little creatures whose numbers have been exploding for some time are suddenly in motion everywhere: an unprecedented reshuffling of the human deck across the entire globe.

(I was certainly doing my part by traveling for more than a year.)

To some extent this is simply because there are more of us than ever before, so more of us are moving. When I was eighteen years old, I traveled outside the US for the first time, moving through a world populated by four billion people. Now there are over seven and a half billion. The number of wild animals and fish in the world fell by half over the same period. Twice as many humans, half as many everything else. Some things are simple.

But the vast human migration now underway is the result of much more than the sheer number of people. War, climate change, poverty, wealth, cheap oil, industrialized agriculture, the internet, "globalization"—I saw it all.

I played three concerts in Lebanon, a country of 4.5 million citizens and 1.5 million refugees. If the US was in a comparable situation, we would be hosting 100 million refugees. In El Salvador, I played in a country of just six million. But there are 2 million more Salvadorans, or one quarter of all Salvadorans, living in the US. I spent two weeks paddling a kayak in the southern Mediterranean, where hundreds of refugees from northern Africa were drowning each week while trying to make it to Europe. Humanitarian rescue boats ran out of deck space on which to place their corpses.

Just within China alone, the greatest migration in human history is now underway. Instead of fleeing war or poverty or climate change, this migration has been thought up and executed by the state. *The plan is for 250 million people to move from the countryside to the cities over the course of twelve years.*

Why in the world would a government want to move 250 million people out of food production and into urban consumption in just over a decade? To prop up their "economic miracle." Chinese leaders want to have a middle class of consumers in place ASAP so that when the economy of the rest of the world tanks or breaks apart, Chinese factories will have someplace to sell their massive output.

All those 250 million people leaving the farm have to go somewhere.

When I made my first concert tour through China four years ago, there were so many skyscrapers under construction that there was a crane shortage in the rest of the world. In the four years between my first trip and this one, China used more cement than the US had used in the entire twentieth century.[5] On that first trip, I rode through southern China for hours on super-fast bullet trains and never had a moment when there was not a group of giant towers under construction within view. Retracing that route in 2015, I still saw more construction than I would have seen anywhere else in the world, but the buildings that were under construction five years ago were now built. In fact, they had already started falling apart. (A common form of corruption in China is skimping on the quality of construction materials.) The families who had moved into them seemed to be falling apart as well, torn from the life on the land they knew and thrust into towering pillars of shoddy workmanship designed with little thought for where human community might go in the new scheme of things. The sight is so grim: armies of drab new towers, grouped into platoons of say a dozen each, the platoons grouped into companies and then whole divisions, brand new conglomerations of identical towers grouped by the dozen, then those groupings subsumed in even larger assemblies, and so on. Entire armies of nearly identical high-rises that continue their grim march to nowhere until abruptly coming to a halt at the edge of rice fields. You look out your window on the forty-first floor and directly into the window of the matching unit on the forty-first floor of the neighboring building. There are traffic jams at the elevators before and after work.

The last attempt by the Chinese Communists at something this crazy, on this scale, was what they called the "Great Leap Forward" in the 1950s. The idea was that China would bypass the whole messy industrial development process and "leap" into superpower status through revolutionary ardor and mass mobilization. Mao convinced the Chinese people to leave their fields and build backyard "steel mills" in which they would transform their ploughs and cooking utensils into steel to make China strong. But no steel was produced. Turns out you actually cannot make steel in your backyard, no matter how revolutionary you feel. Not possible. Thirty million people starved to death. The biggest famine in history, created entirely by government policy.

★

Landing in Jakarta, the capital of Indonesia, the plane dropped me into the center of yet another mass migration from countryside to city. We don't hear about this in the West as much as we hear about China, because we don't hear much about Indonesia at all, despite it being the world's fourth most populous country. But internal migration within Indonesia is now so vast that it too ranks as one of the greatest migrations in history. Jakarta, the capital city, is growing so fast that parts of it are sinking *a foot a year* (one hundred times faster than Venice) because of all the new residents sucking up the groundwater.

The sinking of Jakarta, combined with rising sea levels, increasingly violent storms, and the five hundred tons of trash dumped in the city's waterways each day, has created one of the world's most acute looming catastrophes. The plan, in one of the most corrupt and ineptly managed countries in the world, is to build a $40 billion seawall and seventeen artificial islands so that the city can continue to sink without flooding. Good luck with that.

The home I left behind in San Francisco is in the Mission district. When I moved there in the 1980s, the Mission was mostly Latino, and mostly working-class. Today the Mission is overrun by "techies"—rich, arrogant denizens of the giant Silicon Valley computer companies. These people have a stupendous amount of money. Facebook's Mark Zuckerberg is thirty-three years old and worth $56 billion. Apple is the most valuable company in the world, with over $300 billion in the bank. But even the lowest techies in the food chain make bank. Google hires twenty-two-year-olds just out of public university for a base salary of over $100,000, signing bonuses of $50,000 more, and stock options that total hundreds of thousands more in just a few years.

Wandering around the world, you can see the reason for this extraordinary glut of wealth everywhere you look. From the docks of Makassar to the slums of Lima, from gay discos in Mexico City to lonely backroads in Patagonia, I saw people glued to their iPhones, Google-searching and Facebook-posting. When I left home in 2015, Facebook had 1.4 billion users. When I returned home a little over a year later, there were 300 million more, a number approximately equal to the entire population of the USA.

Facebook did not exist when I moved to the Mission. Neither did Google, Twitter, AirBnB, Uber, and the myriad little startup corporations dreaming of joining the big leagues. Somehow everyone who works for these tech titans wants to live in my neighborhood. Zuckerberg recently bought a house (actually, three houses) just a few blocks from mine. A few blocks the other direction is a house run as a techie hostel, with two bunk beds in each bedroom. The price to sleep in a bunkbed four-to-a-room? $1,200 a month. Living in San Francisco is like living in San Diego/Tijuana. San Francisco may not have an international border, but there are two econo-mies: the tech economy and everyone else. When those who live in the tech economy step outside of it, they are like San Diegans in Tijuana: "Everything is so *cheap* here!"

Is this what living in Detroit was like at the beginning of cars? Does anyone even know what was in Detroit before cars? Not that the early car industry resembled today's tech industry. Detroit had an insatiable appetite for unskilled workers, and a "great migration" of African Americans left the South for a better life in Detroit and the upper Midwest. San Francisco today could not be less interested in unskilled labor. The tech companies scour the world for anyone exceptionally good at math, then do whatever it takes to bring them to San Francisco. The day of the math geek has arrived, dawning so brightly that it is easy to forget that until recently, being good at math did not get you much in this world. An Isaac Newton might do OK, but anything less and you were out of luck. All you could really do was teach, and math departments were academic backwaters. Military technology had no real use for math geeks until the advent of the computer. Neither Clausewitz nor Sun Tzu, the classic military strategists of West and East, thought a great general needed mathematicians.

Of course, the math geeks don't arrive in San Francisco alone. They come with corporate managers, executives, marketing teams, acquisitions and mergers specialists, venture capitalists, human resources people, accountants, tax lawyers, spouses and lovers and kids, and so on—the great tech migration.

The great tech migration merges with the migration of the artistic class (people like me) to a few vibrant cities whose cultural vitality is inflated by the air getting sucked out of everywhere else. Remember that as recently as the 1960s, local music scenes in towns such as St. Louis and Kansas City were sufficiently developed and diverse to each have their own "sound" that was nationally and even internationally known. Today the artistic class migrates to a few magnet cities, so no more Kansas City Sound. Sometimes members of the artistic class will move somewhere like Detroit to try living outside the cash economy, but mostly we go where we can make money from our music, movies, or art. Which means there must be large piles of disposable income sitting around which is not earmarked for necessities like food and housing.

In today's world, lots of cash means techies, so the techies and artists flock together. And not just in San Francisco. From Mexico City to Manila to Montevideo, I circulated in a familiar milieu of intellectuals moving easily between "tech" and "art." In Beijing I had expected this. But I was startled when numerous audience members at a music festival I played at in Manila told me they were working for "internet startups." (The word "startup" has escaped the bounds of English and entered the global lexicon.) I assumed they meant they were telecommuting for companies in faraway places who were taking advantage of the combination of low wages and high English language skills found in the Philippines. But no, they were working for Filipino startups targeting the Filipino market.

Then there is the migration of just about anyone who can afford to leave the suburbs for the few vibrant cities funded by techies and fueled by artists. Their parents or grandparents fled the cities for the suburbs, leaving behind the urban blight of the 1970s and 1980s. In the process they created a cultural wasteland of suburban homogeneity, and their kids are beating a retreat in the opposite direction. Since the increasing polarization and concentration of wealth is a global phenomenon, there are lots of people everywhere who can afford this trip.

The result is gentrification. Americans like to think of themselves as the center of the world, and San Franciscans and New Yorkers like to think the gentrification of their cities is somehow unique. But to think of gentrification as a local phenomenon is to be like the blind men who each touch a different part of an elephant and then argue over what the animal is like. In Shanghai, Berlin, Hanoi, Beirut, Buenos Aires, Istanbul, and Lima, you hear the same conversation about housing prices shooting up to unimaginable levels in the areas where art, tech, and money have attained critical mass, along with wholesale displacement of long-term residents. Doesn't matter if the country you are in is rich or poor, capitalist or communist, North or South, East or West—the only thing that changes is the scale. In Berlin I heard people complain about gentrification that would have hardly warranted comment in the hothouses of New York and San Francisco. But in Asia I saw things that leave New York and San Francisco in the dust.

In Hanoi, housing prices in the hip section of the capital of a poor country in Asia had reached near–San Francisco levels. And yes, there are tech entrepreneurs and startups in that communist capital.

Shanghai may have the largest piles of disposable cash in the world. Shanghai has a twenty-two-year-old billionaire with a dark blue Maserati who spends his Friday nights driving for Uber so he can impress girls with his car.

You can see stuff in China that is so far out on the gentrification scale that we would really need a different word for it. Before moving people into all the new skyscrapers, the state moves in with bulldozers and wrecking balls and knocks their old houses down. I saw people still living in houses from which the street-facing walls had simply been shorn off by bulldozers. What remained looked like a movie set: people trying to live their lives with the "fourth wall" stripped away so that the audience (in this case anyone passing by on the street) could observe the family drama. The debris from the destroyed fourth walls filled the street. Pedestrians had to pick their way carefully through the mess to avoid injury, all the while pretending not to see into the kitchen where the next meal was being prepared, or the bathroom where the last meal was being disposed of.

Picking my way through the concrete rubble of just such a street, I stumbled into a wake for the grandmother of an extended family still residing in a partially destroyed house. When I realized where I was and what they were doing, I made for a partially destroyed door so as not to intrude. But the family insisted I stay. In Chinese culture, they explained, if a stranger stumbles upon such an event it is considered auspicious for the family and for the deceased. So there I passed the afternoon, with kids, grown-ups, elderly, and a group of Buddhist monks in saffron robes and sandals performing the rites, all climbing around and through the rubble of what used to be a house, cooking on a stove in the rubble of what used to be a kitchen, and so on. Not to worry: there was an apartment waiting for them on the X floor of Building Y in Complex Z in the nightmarish skyscraper city down the road.

For the most part I travel as a musician, going to the cities where I play concerts, staying in the hotels the concert organizers have booked for me. But every now and then on this extended trip, I took a break from concerts and dipped my toes into tourism. I discovered a global phenomenon that has exploded so far so fast that, as with "gentrification," "tourism" is no longer an adequate word. What we used to think of as tourism is now better thought of as migration of another sort.

The great tourist migration seems to have really gotten underway about 2010, so if your picture of what tourism looks like is older than that, you are out of date.

Take Sri Lanka for example. In the 1980s about 200,000 tourists came to Sri Lanka every year. By 2002 that number had doubled to 400,000. Ten years later 1,500,000 tourists arrived in one year. In 2016, I was one of some *2,500,000 tourists* to visit the island.

How do those numbers play out in the day to day?

One day I decided to go on a "safari" to Yala National Park, Sri Lanka's principal wildlife refuge, with my friend Ingo who was my traveling companion for a couple of weeks. According to Lonely Planet, "With monkeys crashing through the trees, peacocks in their finest frocks, elephants ambling about and cunning leopards sliding like shadows through the undergrowth, Yala National Park (also known as Ruhunu) is *The Jungle Book* brought to glorious life. This vast region of dry woodland and open patches of grasslands is the big draw of this corner of Sri Lanka. A safari here is well worth all the time, effort, crowds and cost." What did "*Jungle Book* brought to glorious life" actually look like in 2016?

You climb into these giant "jeeps." Ten years ago, there were 10 or 15 jeeps on "safari" in Yala on any given day. The day we went there were 650. Wall to wall jeeps. The sounds of the forest were drowned out by the roar of the jeep behind you, and the smells were overwhelmed by the exhaust belching out of the one in front of you. Then you hit safari gridlock: absolute standstill, drivers shouting and cursing at one another, and a constant blast of exhaust in your face.

You do see the occasional monkey or peacock. But the big prize, according to the guides and drivers, is seeing a leopard.

This is one of the last places in the world with "wild" leopards. Like the snorkeling guide in the Philippines, the Sri Lankan drivers are somewhat mystified by all the tourists. "Tourism" is not part of their culture. Sri Lankans don't go around as tourists in other parts of the world. But the avalanche of tourists coming to Sri Lanka is the livelihood of the guides, and it is a much better livelihood than most Sri Lankans have previously enjoyed. So, just like in the Philippines, keeping these rather odd tourists happy is top priority, which means trying to figure out what tourists want and delivering it.

One thing all guides understand is that tourists want to see a leopard. It is the money shot. So roaring around the reserve we go in search of a leopard. Drivers share leopard tips with other drivers who are their friends and stonewall drivers who are not. We get a tip on a leopard location, and our jeep lurches into high speed. We arrive in a gridlock of jeeps, each one loaded with tourists craning their necks to see a leopard that is allegedly lounging in a bush. The exhaust of all the jeeps is suffocating. Drivers inch forward and back, jockeying for position. Our driver suddenly jumps the road and bypasses the crowd by nearly driving into the bush. Other drivers shout and curse and honk. Our driver and guide pay them no mind but excitedly point. There, through dense foliage, we can see a glimpse what may (or may not) be a piece of leopard fur.

Woohoo. High fives all around. Big smiles. Our driver and guide are exhilarated. They have delivered. Off we go, engine roaring and exhaust spewing. Ingo and I explain that if we inhale any more exhaust we may asphyxiate, and would like the safari to end. On the way out we pass an elephant who is definitely real in the road, foraging in the upper reaches of some trees for lunch. The driver throws it into reverse and hits the gas. Jeeps and elephants sharing the road is not recommended.

Later over beers, Ingo and I gently attempted to explain to our guides that we didn't care whether we saw the leopard and would have been happier just driving around a less-traveled part of the reserve. This was not a statement our guide and driver could even begin to process. They replied that we were tourists and we saw a leopard. Wonderful! A great day! Another beer! Life is good.

In our room later, Ingo and I wondered whether it really was a leopard in the bush or just a piece of leopard fur left for tourists, or even a stuffed animal. We hoped for the latter. If it was an actual leopard, it would be less molested in a zoo, where anyone who crashed over the barrier into the leopard pen in a giant jeep would be arrested.

★

Sri Lanka was one of the places hit hardest by the Southeast Asian tsunami of 2014. Forty thousand Sri Lankans died in a few minutes. The nonstop news coverage put Sri Lanka on the map of the global imagination like never before. The fortunate survivors got lucky twice when the money from a global relief campaign of unprecedented scale poured in. As they will freely say, they "got rich"—at least by Sri Lankan standards. Sri Lankans filed claims reporting losing structures that never existed or damaged boats they never had. Damaged boats that actually existed were passed from neighbor to neighbor so that everyone could get them "replaced." Many used the money to replace humble fishing abodes with humble tourist hotels. Once the tourists showed up, there was more money to be made, much more than you would get working your ass off in the traditional economy. "Local culture just stopped," I was told. Most Sri Lankans, however, do not seem overly concerned about the disruption of their culture. To the contrary. They have more money, do less work, and get to meet people from all over the world.

The 2014 tsunami was limited to Southeast Asia, but the wave of tourism that is inundating remote beaches and mountains is washing over the entire world. It takes about five years to go from almost nothing to complete saturation. In the canals of Kerala in southern India, a few enterprising individuals discovered that the traditional bamboo houseboats used by the locals could be upgraded and marketed to tourists. In five years Kerala's canals went from one of the world's idyllic wonders to houseboat gridlock, with the exhaust the new motors required to move the oversize boats suffocating the fish and other water life. In the remote Filipino fishing town of Coron, it was likewise discovered that their traditional fishing boats—narrow wooden boats with graceful bent-bamboo outriggers—could be upgraded to take tourists on sightseeing trips through the fantastical limestone islands scattered throughout the clear blue water, and in five years the number of fishing boats "upgraded" to tourist-ready jumped from 16 to 160.

The bellwether for all this seems to be Lonely Planet, the ubiquitous book series and website that nearly every tourist uses. The Whole Foods of tourism. Get listed in Lonely Planet and the clock starts ticking. Your little piece of the world will go from pristine to overrun in five years. In the sleepy little Filipino beach town of El Nido, where five years previously there were just a scattered few tourists, so many arrived during the 2016 peak season that they were pitching tents on the basketball court as every single room was booked.

In places like Bali where tourism goes back more than five years, it's tourists-on-tourists. While stuck in one of the endless traffic jams, my Balinese concert organizer pulled the car over to allow police cars with flashing lights to pass, only to discover that they were leading a line of buses filled with Chinese tourists who had apparently paid for police escort just to get through all the other tourists. My host was not at all pleased that Balinese police were forcing Balinese off the road to make room for tourists, but he was not happy with much in Bali. As he explained, traditional Balinese culture—what brought tourists there in the first place—is built on numerous elaborate and extended rituals performed throughout the year, which one must make extensive preparations for by making art or practicing music and dance. These practices developed over the centuries and were tuned to the daily and seasonal rhythms of traditional Balinese agriculture. But now the vast majority of Balinese work in tourism. Forty hours a week. Four weeks a month. There is no downtime after the harvest, no social time when it rains. Tourists still need their hot stone massages and mai tais (a cocktail invented in California) when it is raining. You simply cannot participate in, or ultimately sustain, Balinese culture if you have to be at work forty hours a week.

"Expat communities" have exploded in number alongside the tourism. In fact, the boundary between "tourist" and "expat" is disappearing, as tourists take longer vacations, and expats take advantage of cheap airplane travel to make frequent jaunts back home. An expat is a tourist who stays longer.

An American who had lived in Thailand for twenty years explained how two decades ago, when an American like him came to Thailand and wanted to stay, there were very few options available for how to make that work. Americans arriving in Thailand today can travel until they land somewhere they like, and then stay. But they never really have to *decide* to stay. They can just stay as long as they feel like it. They can return to traveling anytime. It is extraordinary how many Westerners have money to do this. There are the ones in their twenties and just out of college, enjoying a globalized extended adolescence. Then come the techie telecommuters in their thirties and forties, the "early retirees" in their fifties, and finally the retirees who plugged away at their jobs until sixty-five before heading to the beach. Expats are drawn to beaches like flies to flypaper, especially if there is good surfing. They buy property without any idea how long they will stay, or how often they will return. (Many of the newly minted Chinese tourists seem to have even more money than their Western counterparts, but they are far more interested in buying a condo in New York City or London than Chiang Mai or Bali.)

Fueling all of this is a synergistic combination of cheap oil, the global reach of the internet, the global polarization of wealth, and the availability of rich-world-level health care at less-than-rich-world-level prices in otherwise impoverished locations catering to the expat market.

Many years ago my mother got a job as a travel agent when my father was incapacitated. It was a job that one had to train for. I watched her put together exotic beach vacations for her clients. It took a lot of work, more than most Americans back then had time for. It also took real expertise. Different people in her travel agency specialized in different parts of the world, and spent hours sorting out what offers were real, what were scams, what connections were reliable, and so on. Today the travel agent has gone the way of the scribe. Anyone can easily plan a trip to anywhere in the world from their laptop, without doing even the most minimal research into where they are going. I just now performed a random test: in under two minutes I had a list of flights to both Zanzibar and Mongolia, ready to purchase.

Buying a property you have never seen is no more difficult than buying the plane ticket. Just click. Then stay in touch with friends and family back home via social media. One expat told me she communicated *more* with her friends and family back home from Sri Lanka than she did when she lived in their neighborhood. She has lots of free time now, and she spends most of it chatting with people on the other side of the world. She has not even taken the first steps toward learning Sinhalese or Tamil. She knows very little of Sri Lankan culture, though she seems quite knowledgeable about Sri Lankan real estate prices and tax law.

Many people can telecommute back home to work. If your job involves sitting in front of a networked screen, you can sit in front of a networked screen almost anywhere. If the main way people at your company communicate is via networked screens, your coworkers may not even notice you are gone. Maybe you have to take a cut in pay for not working on site, but that is more than offset by the lower cost of living in your exotic new home. You will be living better making less. Which means you have more time to surf.

I met a young woman in Bali who was working for an environmental NGO preparing to launch a Balinese campaign. Her job was to visit the villages where the campaign hoped to have a presence, and identify local leaders who the campaign would then reach out to. She was following a set of techniques frequently used by community organizers in the US. She was not looking for elected leaders or those with the loudest voices but for the people others turn to for problem solving and advice. The ones who, when they sign on to a campaign, their neighbors follow. Time and again, her path led her to expat surfers, which made a certain kind of sense. They have the most money. They have traveled widely. They have a larger-than-village experience and world perspective. And, hey, they are hanging out surfing while everyone else is working, so they must know something everyone else doesn't.

San Juan del Sur was a dusty little beach town in Nicaragua when I first visited in 1980, so isolated that the beach was used for smuggling weapons during the Central American civil wars of the 1980s. No one even noticed the surf. After the wars, the waves were discovered. Today, San Juan del Sur is overrun with ugly condos and luxury homes that sell for near-San Francisco prices, in the second-poorest country in the Western Hemisphere. There is a local newspaper—in English—which mostly covers real estate for expats from the US, Canada, and Europe. Several seasons of the hit reality TV show *Survivor* were shot there.

I played concerts in Chiang Mai (Thailand) and Yogyakarta (Indonesia) that were organized in part by people from Western countries who have lived in Chiang Mai and

Yogyakarta for twenty years. They had married local spouses, became fluent in the local language, and integrated into the local economy to the point that they could longer afford to visit their home countries. The expats of today never really have to get their hands dirty with the local culture. You see this all over the world: expats who live entirely in a bubble. Who don't know any locals. Who can fly in and out whenever they feel like it. Who will live somewhere for years and never learn a word of the local language. They are not interested. They are not there for the local culture. They come for the weather, the high quality health care tailored exclusively for them and unavailable to the local population, and the servants they can afford while spending their first-world pensions at the third-world beach.

There is no way to travel the world and not be part of the tourism wave, just as there is no way to open an art gallery in a low-income part of a city—any city—and not be part of the gentrification wave. You can squirm and fuss all you want to about how you are an "artist" and therefore what you are doing is somehow not part of all of that, but this is delusion. "Gentrification," if you want to call it that, takes the form of rising rents in San Francisco, bulldozed homes in China, tourism in Sri Lanka, and expats in Nicaragua. We are all just drops of water in the tsunami of migration that is redistributing the global human population.

In the Philippine waters off Coron Island, I paddled a kayak past numerous large resorts under construction, as well as the fourth cruise ship ever to ply those waters. That is where my drop in the wave is: I show up about the time the fourth cruise ship arrives, after the construction of the big resorts begins but before it is finished.

★

Coron Island is a sixty-square-mile jewel in the warm Pacific waters in the northern reaches of the Palawan archipelago, itself a subset of the Philippine archipelago, which is in turn a subset of the Malay archipelago. It is also the most beautiful place I have ever been: tropical forests, crystal clear lakes, lovely beaches, and, most dramatically, towering limestone cliffs. When I show pictures of Coron Island to friends, they almost invariably ask, "Is that *real?*" It is hard to imagine that the place in the image exists in the real world and not in Photoshop.

In 1998 the Filipino government designated Coron Island the ancestral homeland of the Tagbanwa people, one of the few surviving indigenous peoples of the Philippines. The Tagbanwa on Coron are a distinct tribe among the Tagbanwa, and there are only about two thousand of them on earth, all on this remote island.

For as far back as anyone can remember, the Tagbanwa of Coron have lived in extreme poverty. But not long after their island was designated their ancestral homeland, tourism in the area began to take off, and the fishermen of the nearby town of Coron on Palawan Island converted those 160 hand-made bamboo-outrigger fishing boats into sightseeing vessels for tourists. The Tagbanwa who owned coastal land discovered they could charge tourists for landing on the beaches of Coron Island. This was a stunning reversal of fortune, because nothing can be grown on sand, so before the tourists showed up the beaches were worthless. Now the beach owners are the wealthiest Tagbanwa on the island, and they don't have to work. They just sit on the beaches waiting to collect money from tourists.

Before the tourist money arrived, the Tagbanwa were at the bottom of the social ladder in the region, the butt of every racist joke. Now young Tagbanwa men take their money into the town of Coron and blow it on wild-west nights of strong alcohol and loose women.

I tagged along with a tourist couple from Slovakia on an overnight beach camping trip to Coron Island they had arranged with one of the many converted fishing boats. The beach was stunning, dotted with delicate limestone sculpted by time in ways no human mind could imagine. The guides brought us a delicious dinner of fish they had caught minutes before and grilled on the boat. The men of the Tagbanwa family that owned the beach were camped out on a bamboo platform at the far end of the beach, waiting to collect fees from whoever might arrive. When the Tagbanwa first started collecting beach fees, not that long ago, they asked 25 pesos for a night (something like five American cents). Now they were asking 1,500. Our guide was shocked, and bargained them down from 1,500 pesos to 300, but the Slovakians refused to pay even that. They were nice enough people I am sure, but, like so many tourists, they seemed less focused on the unique beauty of the culture and landscape they were visiting than on nickel-and-diming every local they came into contact with. The central goal of their vacation seemed to be avoiding getting "cheated" by locals. In this they were no different than many—perhaps the majority—of tourists in the area. They were convinced their snorkeling guide was overcharging them until I pointed out the home-made wooden flippers on his feet, and asked if they really thought a professional snorkel guide who could not afford to buy commercially manufactured flippers was raking it in.

I spent the next day paddling a kayak around Coron Island on my own. While paddling past a beach I noticed a local boat guide from Coron town on shore waving me toward him. He seemed to be fascinated by my kayak. I landed and engaged in friendly paddler-guy chitchat about kayaks and outriggers and boat design and whatnot, until a Tagbanwa woman appeared to demand a fee for landing my kayak on the beach. Using the guide as an interpreter, I explained that I had no cash with me, that I had not intended on landing on the beach until the guide waved me in, and that I would leave immediately. The smile that had graced her face when asking for money did not change in the least, but in a tone that conveyed the exact opposite of her countenance she ordered me, in English, *"Get out."*

I scurried away, confused and unnerved. This was not what I had imagined my first encounter with indigenous Southeast Asian culture would be.

So, how much *should* it cost to for a tourist to land on the beach of a beautiful limestone island that is the ancestral homeland of a small group of indigenous people? This question filled my head as I paddled away. I decided that if the answer was up to me, the cost of landing on the beach would be one million US dollars. At that price, once a year someone like that twenty-two-year-old billionaire in Beijing who spends Friday nights driving his Maserati for Uber to impress girls would bring a date to Coron Island just to show off his wealth. Neither he nor his date would even be interested in the place and would leave quickly. The Tagbanwa would far make more money per year than they do now, avoid the tsunami of tourists that is just a year or two away from washing up on their shores, along with the environmental devastation that will surely accompany it, and go on with their lives. Or rather, go on with whatever their lives will be with a yearly influx of a million dollars.

Go on with their lives, that is, if they can get off their smartphones. Because these days, when you land a boat on a Coron Island beach, as often as not you will find the folks waiting there to collect your landing fee engrossed in playing World of Warcraft on their smartphones.

Smartphone World

Remember the "digital divide" that caused such worry just a short while ago? The idea that the divide between the haves and the have-nots of the world would be exacerbated by a new divide between those who have computers and those who don't?

Not a problem.*

The opposite is true. Computers, in the form of smartphones, have become the most widely used technology in the world. More than television. More than cars. More than indoor plumbing. More than roads.

These little computers are changing how we communicate, what we communicate, with whom we communicate, and with whom we don't communicate. They are changing not just how we spend our time but also our very sense of time. They are changing who people are at a deep level, and they are doing so fast.

Take something as trivial as what people do while attending concerts. The concerts I play range from abstract electronic music to virtuosic contemporary compositions. Until smartphones became ubiquitous, people attending my concerts listened with undivided attention and fierce intensity. This had nothing to do with, say, the formality of a symphony hall. It was simply the experience that people came for, that they treasured.

Most of my concerts are in Europe. Today, when I play in Europe, pretty much everyone under thirty in attendance is on their smartphone throughout the show: texting, WhatsApping, Facebooking, Instagramming, Grindring, or taking pictures or videos. Which is funny, because there is nothing to take a picture of. I generally stand stock still when I play. The sound is the whole point. To someone like me, it seems that the people

---

* The Covid-19 pandemic has brought renewed attention to the idea of a digital divide, as social distancing pushed education from the classroom to the internet. According to the US Department of Education, 14 percent of children between the ages of three and eight did not have internet at home in 2017. In a country that aspires to provide universal public education (if the US can still be said to do so in today), leaving 14 percent of schoolkids out of the virtual classroom is a major problem and indeed a digital divide. In this essay, I use "digital divide" to refer to something quite different: not the percentage of Americans who have a dedicated internet service in their homes, but the percentage of people in the world who have access to the internet at all.

on their phones are not paying much attention to the music. But then after the concerts, when I stop playing, they put their phones away and gather around the stage to tell me how much they liked the music. Weird. But that's Europe.

In Shanghai I actually cut a performance short, leaving the stage after maybe ten or fifteen minutes. The people gathered around the bar tables in the club did not even look up from the phones when I started playing. Many seemed as oblivious to the abrupt end of the performance as they had been to the beginning. The concert organizers rushed to find me backstage, looking baffled and concerned. Was I OK? Was there a problem? There was a major understanding gap between us, and it went deeper than the language barrier between English and Chinese. The barrier was not linguistic but technological. As the tour progressed, I learned that this is just the way concerts are in China. It troubled many Chinese musicians as much or more as it troubled me. The only time I saw Chinese people attend a concert without texting throughout the show was when a Chinese clarinetist ingeniously incorporated everyone's smartphones into his composition. He passed out a paper with five simple instructions which allowed everyone to operate their phone as an echo device for his clarinet. Everyone held up their phones, recording the sound in the room, and echoing it back as they chose. Altogether it created a beautiful soundscape, but the principal achievement was to get everyone to listen intently, not by putting their phones away (impossible) but by incorporating the phones into the performance. If you can't beat 'em, join 'em.

OK, but that is China after all. The place where they actually make iPhones and pretty much every other kind smartphone.

But in the mountains of El Salvador I saw the same thing. In the most remote mountains in one of the poorest countries in the world, I attended an intimate night of folk music in a dirt poor community of former war refugees. Local musicians took turns at the mic alongside visiting campesino luminaries from neighboring Nicaragua. It was the big musical event of the year, a few minutes of amateur stardom. Much of the audience texted right through every act.

Raja Ampat, the Papuan archipelago where I went scuba diving and kayaking, may be said with some justification to be one of the most remote places on earth. Just 50,000 people spread across over 1,500 islands. Languages differ village by village and island by island, so many languages that linguists group them into nine separate language groups. The locals live in crude beach houses, largely outside the cash economy. They subsist on the fish they catch, a few meager vegetables they plant, wild bananas, and rice they purchase. They have very little in the way of worldly possessions. Children run naked. Education of the sort we think of in the US is nonexistent.

At the first homestay, or bamboo hut, my guide Sedik and I paddled our kayaks to in Raja Ampat, we found the proprietor alone on the beach, a smartphone in each hand (each phone used a different cellular network which reached different islands). Later that night, after a dinner of rice, fish, and water, I went to check out the beach hammock stretched between two palm trees, only to discover the proprietor's two young sons wrapped inside, each playing games on his own smartphone.

A few days later Sedik and I paddled up to an even cruder homestay and found the proprietor, Metos, chatting with a middle-aged woman from Amsterdam. Metos barely had access to fresh water. There was enough for us to drink if we were frugal, but not to wash with. He had nothing to eat other than rice and fish, not even a banana. But he had a smartphone and a website. The Dutch woman had booked Metos's bamboo hut through his website, then flown from Amsterdam to Jakarta, changed to a local airline to fly to Sorong, and taken the ferry to Waisai, where Metos had picked her up in his rickety motorboat, and after several hours of travel, here she was. With his smartphone and website, Metos can talk to Dutch people in Amsterdam (he speaks basic internet English), but he cannot talk to the people on the next island because they speak a different language.

My time in the Papua region was spent offshore in the Raja Ampat archipelago, so I did not visit the inland forests of the main island. But I was told that, had I ventured into the highlands of Papua, I would have seen some of the last men in the world who have never left the forest, walking around wearing nothing but a penis gourd held in place by a string tied around their testicles, carrying traditional mesh bags and iPhones.

In Sri Lanka, I saw people use smartphone minutes as currency. If I owe you money in Sri Lanka, we can just take out our phones, poke some virtual buttons, and transfer the equivalent of the money I owe, in phone minutes, from me to you. Even some shops accept payment in minutes. You can go to the butcher and buy a chicken with smartphone minutes. Then the butcher can take his minutes to the phone kiosk down the street and trade them for cash. This is not Bitcoin, which is a virtual currency intentionally developed by engineers. Sri Lankan phone time currency simply emerged by happenstance from the widespread use of smartphones. The Sri Lankan government is trying to shut this all down, because this is liquidity in the financial system which the state cannot regulate. Good luck with that.

Sri Lanka is the only place I saw smartphone-minute-currency, but everywhere I went I found poor people whose livelihoods are as dependent on social media as Mark Zuckerberg's. Ori is a poor young man eking out an existence in Java using his phone. As the eldest son in a family in which the father is laboring for starvation wages on a palm oil plantation, Ori is responsible for his entire extended family, who live together in a single room with no bathroom or even a real kitchen. The entirety of their worldly possessions consisted of their clothes, a small gas burner, a motorbike, and a smartphone.

Here is the business Ori figured out to keep the family fed. Every night his mother stays up all night cooking on the gas burner while the family sleeps on the floor around her. At 6:00 am Ori wakes up and uses his smartphone to take pictures of the food his mother cooked. Then he posts the pics on Instagram. Ori's customers choose what they want by looking at the Instagram pics. His mother goes to sleep and Ori spends the day delivering the food on his motorbike. At 6:00 pm he returns home, wakes up his mother, and takes her to the night market to buy ingredients with the money he made delivering the food. Then Ori and his family go to sleep while his mother stays up cooking, and the cycle repeats. Ori also has a Facebook page for people to give testimonials regarding his mother's Sumatran cooking. This is a family of eight in a single room with no bathroom or kitchen, and barely enough money to pay for their next meal. Their smartphone and the social media presence that it brings mark the precarious border between having enough to eat and hunger.

Ori dreams about visiting San Francisco. "If you could go, what would you like to see," I asked. "Facebook and Instagram," he replied. He was visibly startled when I told him that all there was to see were corporate office buildings like any others.

Momo is twenty years old and living hand to mouth in Yogyakarta, Indonesia. He has been on his own since he was sixteen. In the narrative of the "digital divide" between the world's haves and have-nots, Momo should have been a poster child of who was going to be on the losing side: a street kid in a third-world country without stable income or housing, and no obvious prospects for either. Yet since he set out from home, Momo has made his living from smartphones. He buys minutes in bulk and resells them in smaller bundles at a markup. He keeps a careful eye on new technologies in the smartphone market, and buys used phones for cheap that he calculates he can use for a few months and then resell at a profit. Momo doesn't make a lot of money from this. He lives in a boxlike room with no windows, no ventilation, and no space for anything more than his bed. His diet consists of the cheapest junk food from convenience stores. But he is alive, and better off than other street kids his age who are not as clued into smartphones as Momo.

Momo says his smartphone provides him with more than a livelihood. It transforms his windowless room from a jail into a home. When he is shut in there alone he has the whole virtual world to himself. He used to watch movies and YouTube, but he has figured out that playing games uses much less data, so no more videos for Momo. He has entirely committed to games. Western parents worry that their children disappear for hours into a virtual world that cuts them off from the real world. That is precisely why Momo loves his phone: the all-encompassing way it transports him out of his miserable real world. Add to that the fact that he makes his meagre living from it, and the possibility that if he works very hard at learning everything he can about the phone, there is a credible chance that this knowledge might actually lead him out of poverty. There is no other poverty escape route in his life.

Momo told me that his ambition in life was "to become as smart as my phone."

"Are you really so sure your phone is all that smart?"

"Oh, come on, Bob. If I was as smart as my phone I could speak every language."

For decades, computer scientists and philosophers debated whether computers would acquire intelligence. Sides were chosen. Books written. Careers made. But intelligence turned out to be a slippery notion, the debate floundered in arguments about definitions. All agreed, however, that there were activities that certainly required intelligence, such as playing chess. So if a computer could beat the best human chess player, that would be a big win for the computers-can-be-intelligent team.

The "no" partisans confidently declared that no computer would ever accomplish this feat. Then an IBM computer defeated the human chess champion, and the "no" camp reconsidered whether chess was such a good measure of intelligence after all. Since then, the goal post has moved from chess to natural language. (A natural language is one that evolved over time through widespread use, like Swahili or German, rather than invented, like Esperanto or C++.) Understanding natural language requires understanding context, nuance, metaphor, sarcasm, and irony—none of which are involved in chess. So the challenge for artificial intelligence moved from chess to *Jeopardy*, the TV game show in which the contestants are given an answer and must guess the question. This requires understanding context, nuance, metaphor, sarcasm, and irony. Then in 2011 a computer beat the human world champion Jeopardy player. A few months later Apple introduced Siri, and computers that can be addressed with natural language escaped the bounds of the lab and entered our pockets. Even Momo's pocket.

Momo has taken sides in this debate. A street kid in Yogyakarta who never attended high school, he has made an intellectual leap from his stifling windowless room to the boardrooms and labs of Silicon Valley, and precisely identified the crux of the matter. As far as he is concerned, a phone which can translate natural languages is intelligent. Becoming intelligent like that is his lifetime dream.

I encountered this again and again: an uncanny grasp of the cutting edge issues concerning computers and philosophy, by poor people in poor parts of the world, who skipped rotary phones and pushbutton phones and cordless phones and flip phones and pagers and desktops and laptops and tablets, and went straight to smartphones. Ori, who grew up in a village that was a five-hour dirt bike ride from electrical power, went straight from no electricity to smartphones. These people take smartphones at interface value. They are not troubled by ethical or existential questions. Smartphones don't threaten their humanity. The technology makes the daily struggle for survival easier and makes living in a box without windows or ventilation more bearable. The devices come from faraway places and are the result of faraway processes over which these people have no influence whatsoever. I sense that these people are the future, in the same sense that Sedik, clearing the palm branches off the beach while not noticing the plastic garbage, and the homestay proprietor decorating his bamboo huts with seashells and plastic garbage, are the future.

It is worthwhile to listen carefully to what they have to say, for these people will never again exist. Already their own children are born into the digital world. The thoughts and hopes and fears of predigital peoples encountering computers for the first time must be recorded now or never. They are the last window into who humans were before computers.

Grindr World

I had my own experience with the phone intelligence that fascinates Momo as I laid in bed next to Christian, a man from the Maluku Islands. Part of Indonesia, the Maluku Islands were the original "Spice Islands," the only place in the world to which the coveted nutmeg, clove, and mace were indigenous. The Malukus host some 130 languages, and Christian does not speak fluent Indonesian, let alone English. But we were able to engage in pillow talk by placing a smartphone between us running the Google Translate app. I spoke in English and he in Indonesian, and the app would detect which language was spoken and repeat the phrase aloud in the other language. Of course, the app made lots of mistakes, just as any human not yet fluent in a language would. But it was sufficiently accurate for Christian and I to meaningfully converse. Google will keep working on that language and many others, 24/7, 365 days a year, something no human learning a language will do.

This was twenty-first century techno-intimacy. Here we were, in bed, engaging in pillow talk via a computer smaller than a deck of cards lying between us. And we had met through Grindr, the gay hookup app. I live in one of the most expensive, high-tech, tourist-attracting cities in the world. Christian is from a poor, sparsely populated Southeast Asian island where almost no one goes who doesn't live there. Christian had never seen a flushing toilet. I found him in the hotel room bathroom trying to flush the toilet by pouring water from the sink faucet into the ice bucket, then from the ice bucket into the toilet bowl. He jumped with a start when I showed him the toilet handle and flushed. Yet here he was, with a smartphone, all Grindred up, and running social media accounts on Facebook and Instagram.

Bert was another lovely man I met on Grindr, this time in the Philippines. We made beautiful love in a hotel room that opened on to a beach bathed in moonlight. The moment the fireworks were done, Bert whipped out his smartphone in flashlight mode to make a close examination of what white-people cum looked like. His curiosity satisfied, he rolled over and disappeared into his phone.

He stayed disappeared into his phone for a long time. A very long time.

I tried to take it all in. I was on a beautiful tropical beach. The Southeast Asian moon shone down on our naked bodies, which glistened so differently in their different shadings. The Southeast Asian surf gently lapped at the beach just outside our open door. We had just made love. He was on his smartphone.

Time passed.

I was flummoxed. Eventually I took out the iPad which I use to write my journal, and wrote about Bert and his phone. There we were, side by side, naked in bed, each at our separate screen. We could have been anywhere. New Jersey, say.

Finally Bert became curious about what I was doing on my screen. I showed him. He was embarrassed to find I was writing about him and his phone. He put his phone away and, for the first time, we talked. About his phone.

He had been playing Clash of Clans. He plays constantly. He is part of a "clan" of real people scattered across the real world who play together virtually. Since they live in an assortment of time zones they are always playing 24/7. The sun is always shining somewhere.

Bert had gotten his first smartphone three months before, when he came to the beach to work in a small hotel. Since then, his social life has existed almost entirely on his phone. He got a Grindr account right away, but I was his first Grindr date. In fact, he says I am the first person he has had sex with. There is not a lot of Grindr action in this village. Clash of Clans, however, is always there. His real-world friends tell him he is a phone addict. He replies that if he is an addict then his uncle is an even bigger addict. His uncle plays Clash of Clans while driving.

★

Grindr launched in 2008, just seven years before I left home. By the time I met Bert, it drew three million users a day. Grindr's press kit claims the app is active in 196 countries. There are 196 countries in the world (counting Taiwan).

In countries that already had an "out" gay culture, Grindr is sucking the wind out of gay nightlife. Gay bars are empty because everyone is at home staring at their phone. In Nuremberg I discovered there is not a single gay bar left in the city, despite the fact that there seem to be more gay men living in Nuremberg than ever—on their phones. It is common to meet young gay men in San Francisco or New York who have never had sex that was not arranged through Grindr. The skill set required to negotiate transforming a social encounter into a sexual encounter—what used to be called "cruising"—is disappearing. Young gay men still go out, usually for dancing, but if they want to get laid, they go home and stare at their phones. Grindr "dates" can lead to sex of the most depressing, empty sort. Men obsessively swipe through the profiles without ever actually meeting anyone because of an unspoken fear of how vapid an actual encounter would be. Grindr "flaking"—making plans to meet and then simply not showing up—has become as common as actually following through on plans. Those of us who remember the pre-Grindr gay world experience this as a loss.

But for men in parts of the world where social taboos make "out" gay life impossible, Grindr is delivering their gay liberation moment—albeit a technologically mediated one. Men can meet each other safely for the first time. (Relatively safely, actually: the question of whether police are monitoring Grindr looms large in countries where homosexuality is heavily taboo.) Men can also get a sense of how many others like them there are in their village, city, region, and country. Guys in these places don't spend hours, days, weeks, years swiping through Grindr profiles without meeting anyone. The technology is actually delivering on its promise of bringing people closer together. There is a wonderful feeling of celebration and empowerment about it all. For someone like me, traveling the world far off the gay tourism path, Grindr provides an effective escape route from the tourist bubble: arrive someplace new and go on Grindr, and quickly you can make local friends, see their homes, even meet their families.

This may all change soon, as the techno-dysfunction of first-world Grindr-ism goes global. The leading edges of this are all too obvious, especially in places like Beirut, where closeted gay men from around the Arab world go for Grindr escapades amped up by party drugs and testosterone. Grindr traveling also reveals the stunning homogenization of techno-social norms Grindr brings in its wake. Grindr has made gay culture truly global. The Grindr earth is flat. Grindr is now, hands down, the way men on earth learn how to be gay, what it means to be gay. No matter their native tongue or language skills, everyone on Grindr everywhere in the world knows Grindr English.

Shifts in Grindr culture are propagated globally at internet speed. Not long before I left home at the start of my trip, a man on Grindr in San Francisco sent me a close-up pic of his asshole as a way of propositioning me. This was a first. I was appalled. "Poor guy," I thought. "Someone needs to tell him that this is not really the idea." Within a few weeks I was regularly receiving asshole close-ups. Grindr asshole close-ups had gone global. Traveling around the world, my phone has received close-up pics of Indonesian assholes, Mexican assholes, Malaysian assholes, Cambodian assholes, Peruvian assholes, Italian assholes, Portuguese assholes, German assholes, Armenian assholes, and American assholes.

Question: In the age of social media, how fast can a social activity go from never-existed to a globally understood code?

Answer: a few weeks.

Until the last year, how many close-up pictures of assholes were there in the entire world? Now there are thousands. Will they too make it into the geological record?

A few months after I returned home, a friend I had met on Grindr in a part of the world where homosexuality is strictly forbidden came to visit me in San Francisco. I took him for a walk through a neighborhood with many gay bars. He was incredulous. In broken English, he blurted out, "It is like live Grindr."

Selfie World

All those smartphones have cameras, of course. Remember all those scenes of tourists from rich countries walking around dusty third world villages intrusively snapping pics of the colorful local culture? Surely you have seen them. You might have been in some of those scenes yourself. They have been replaced by scenes of wealthy tourists from rich countries walking around dusty third world villages snapping pics of the locals—who are busy snapping their own pics of the colorful foreigners. Sometimes outdueling the tourists in pics-per-minute. On several occasions I witnessed tourists beat a hasty retreat from places they had traveled long and far to see, overwhelmed by a swarm of locals clamoring for one, two, many selfies.

In Makassar, a port city on the island of Sulawesi, a man told me he had three urgent questions for the first American he had ever met: Was I going to vote for Donald Trump? Did Americans understand that all Muslims are not terrorists? Can you buy selfie sticks in the USA?

Like the smartphone, the "selfie" (the idea, the act, and even the word) is a global phenomenon. Even more global than McDonalds, Nike, or Coke. Unlike those global brands, which bring with them ideas of how to eat, dress, or drink, the selfie brings with it a way to *be*. If men the world over are taking their clues on how to be gay from Grindr, the selfie is teaching everyone everywhere how to perform happiness in the same way. No matter where you are in the world, no matter your race or creed or caste, and no matter how you are feeling, selfie time means dropping everything and *performing happiness*. And the standards for an acceptable performance are the same everywhere. The same finite set of facial looks, hand gestures, postures, leaps into the air, and so on.

How new is the selfie? Susan Sontag's seminal 1977 book *On Photography*, which is now celebrated for having been so prescient about the impact photography would come to have on culture, did not mention the act of taking a photograph of oneself as anything anyone had ever done or would ever do. It simply didn't occur to Sontag.

I watched a group of Indonesians in their early twenties arrive on a beach at the southernmost tip of Sulawesi and immediately begin performing happiness for their smartphone cameras. They smiled. They posed. They organized to get shots of themselves running down the beach with their hair blowing in the wind. They regrouped again to organize shots of themselves running euphorically into the water, coordinating the motions of the body-through-water and finger-on-button to capture maximum splash. Once they had the pic, they all turned around and came right back out of the water. They were acting out the Western beachgoers' ritual that they had seen countless times on screens but without the dive and swim that was the point. The selfie was the point. Once you had it there was no reason to stay in the water. One by one, they got selfies of themselves performing every selfie trope on the list. Watching them, you realize what a defined list it is. It is a new global code for happiness. The global code even specifies saying "one, two, three" in English before hitting the button, no matter where in the world you are or what language you speak.

But they were not just going through the motions of happiness. To the contrary, they were delighted. Radiant. Not faking it. The happiness was real. But it was the image of the sand and sun and water and their bodies, not the touch of the real sand and sun and water on their bodies, that made them so happy. People where I live notice that. It bothers us. On a beach in my country, chances are high that at some point one of these beachgoers would say, "Hey, you guys, this is weird. Let's put the phones down and just swim." We complain about how nobody notices when we get sucked out of the world of sand and sun and water and into the world of images of sand and sun and water, but the fact that we complain about it at all shows that we do notice, and it bothers us. We take this kind of critical distance from the technology as a cultural given, and it is disconcerting to be in a place where this distance is absent.

Accepting the image of sand as sand is like accepting phone intelligence as intelligence, or accepting bits of colorful plastic floating in from the sea just as you accept the colorful things that grow there. This is not my culture. This is the culture of people whose lives were not part of the evolution of this technology, but for whom this technology fell from the sky, or was unloaded from a boat, or washed up on the beach. The odd thing is that I live near Silicon Valley, the place where all of this started, but it was the kids at the beach in Sulawesi who are revealing the future to me.

★

At the ancient Buddhist monument of Borobudur, I watched Buddhist monks in saffron robes taking selfies of themselves praying.

Buddha taught that the self is an illusion.

A Zen koan for the new century: when a Buddhist monk takes a selfie, what is in the picture?

Photo credit: Bob Ostertag

Postscript: 2020

Four years have past since that sunny afternoon in the Berlin park where I contemplated the world I had traversed during the previous fifteen months. It seemed to be fully saturated with social media. But in just the three years I have been home, an additional three quarters of a billion people have begun posting, sharing, and liking on Facebook. That is equivalent to the entire populations of Indonesia, Pakistan, and Brazil (the fourth, fifth, and sixth most populous countries in the world). It works out to about twenty million new Facebookers per month. One third of all human beings now check their Facebook at least once a month. The number of Grindr users has increased by the same proportions over the same years.[6]

In Java, Ori has grown his food delivery microbusiness into a stable, Facebook-and-Instagram-based business that employs not only his extended family but also twenty other villagers who have come to the city to work for him. The village boy who grew up a five-hour drive from electricity now has ten thousand Instagram followers. He traveled from Java to San Francisco to visit me. One day we had no plans. I told him we could do anything he liked. He wanted to go to Silicon Valley, where he made a video selfie standing in front of Facebook corporate headquarters saying, "Thank you, Facebook, for helping me feed by family."

For many people, however, Facebook is seen in increasingly dark terms, blamed in varying degrees for the ascendance of Donald Trump in the US, Boris Johnson in the UK, dysfunctionally divisive politics the world over, lynchings in India, religious pogroms in Sri Lanka, and genocide in Myanmar.[7] In both India and Sri Lanka the state has shut down social media for extended periods to prevent mass murder.[8] Even Republicans in Washington, DC, increasingly speak of breaking up Facebook, or regulating it, or *doing something*, though they don't know what. Which is likely why Mark Zuckerberg holds secret dinners with Donald Trump at the White House.[9]

As if the amount of time we voluntarily spend looking at the screens of our smartphones was not enough, police in Hong Kong want more. During the massive protests of 2019, riot police made a practice of forcing protestors to look at their phones, even prying their eyes open as they lay on the street unconscious, so that facial recognition algorithms would unlock their phones and enable police to search the contents. Looking away was not an option.[10]

★

I have had the privilege of spending most of the last few years at home. Not so for those fleeing the ecological and political catastrophes I observed during my journey. One out of every 155 Hondurans crossed the US-Mexico border in just one five-month period.[11] In Guatemala's presidential election in 2019, fewer Guatemalans voted for the winner than had been apprehended while trying to enter the United States during the previous eight months.

The Detroit water crisis has been eclipsed by a water crisis of far greater magnitude in nearby Flint. The same Republican governor of Michigan who sent the emergency financial manager to lord over Detroit also sent emergency financial managers to Flint, where the collapse of the manufacturing base and subsequent population implosion led that city too into bankruptcy. But while the viceroy of Detroit sought to balance the books on the backs of the poor by collecting old and often outlandish water bills, viceroys in Flint changed the source of the entire city's water to save money. Almost immediately, Flint residents began complaining that their water smelled terrible, tasted like metal, and gave them skin rashes. They confronted elected officials only to be told, again and again, that the water was fine. In fact, it was contaminated with lead and copper.[12]

Fifteen public officials, including two emergency managers, have been criminally charged and face prison sentences up to a combined 176 years.[13] None have yet been tried. Nearly thirty thousand schoolchildren have been exposed to a neurotoxin known to have detrimental effects on children's developing brains and nervous systems. Twenty-eight percent of Flint's schoolchildren now qualify for special education services, and it is anticipated that they will require special assistance throughout their entire lives.[14] It would be difficult to fund lifelong services on that scale in the wealthiest zip codes in the nation. Flint is one of the poorest.

The air in Hanoi is now even worse than when I was there. Hanoi often has the worst air of any city in the world. If Hanoi is not at the top of the polluted air list on any given day, Jakarta might take that honor instead.[15]

In the San Francisco Bay area where I live, wildfires fueled by climate change have filled the air with smoke for days on end, several summers in a row. Authorities announce changes in the air quality by the hour, along with warnings of who should stay indoors, who can be outside and for how long, and what sort of mask to wear. But our worst air days would pass without comment in Jakarta and Hanoi.

Traffic in Jakarta is now so bad that gridlock alone accounts for $7 billion in economic losses per year.[16] But Jakarta's traffic problem pales in comparison to its sinking problem. The most recent studies predict that the homes of all thirty million residents will be flooded by 2050, yet two hundred thousand more people continue to arrive each year. The plan to build a $40 billion seawall and seventeen artificial islands to protect the city was scrapped even though some sections had already been built. In December 2019, one of those sections collapsed.[17] A more modest plan for building a forty-kilometer seawall at a tenth of the cost has been announced, but even the beginning of actual construction is years away. The Indonesian government meanwhile announced plans to build an entirely new capital city elsewhere, on the island of Kalimantan, at an estimated cost of $32.7 billion.

The rivers running through Jakarta now discharge more than two thousand tons of plastic into the sea each year, which is not even one thousandth of the 3.2 million tons of plastic the ocean receives each year from Indonesia as a whole.[18] The plastic carpeting the beaches in remote Raja Ampat seemed to bother almost no one but me, but in 2017 massive waves of plastic garbage began overwhelming the beaches most popular with foreign surfers in Bali, prompting the Indonesia government to declare a "garbage emergency." Work crews were sent to attempt what I failed to accomplish in Raja Ampat: clean the beaches of plastic. They carted off one hundred tons of plastic garbage each day.

The fact of the global plastic crisis smacked me in the face again and again in 2016. But the scale of the crisis is only just now coming into view. In 2019, researchers were shocked to find microplastic in the rain falling on cities and remote mountain tops, in Arctic snow, and in ocean waters.[19] California's famed Monterey Bay was found to contain concentrations of plastic at a depth of two hundred meters as high as at the surface in the infamous Great Pacific Garbage Patch.[20] Researchers were still finding plastic at a depth of one kilometer. They have been unable to look below that, but there is every reason to think that the plastic continues all the way down and then accumulates on the ocean floor.[21] What researchers have not found is any place on land or sea free of microplastic.

So, silly me for thinking I had to travel far to find a sea filled with plastic.

★

Just after I returned home in 2016, the Great Barrier Reef suffered unprecedented back-to-back mass bleaching events caused by record-breaking marine heat waves. Huge sections of the largest reef system in the world died, and two thirds of the entire system bleached.

In 2019, *Nature* published a study showing that the reef's reproductive capacity had declined 89 percent since that bleaching, concluding that "the extent to which the Great Barrier Reef will be able to recover from the collapse . . . remains uncertain.[22]

Outside magazine published an obituary:

The Great Barrier Reef of Australia passed away . . . after a long illness. It was 25 million years old. For most of its life, the reef was the world's largest living structure,

and the only one visible from space. It was 1,400 miles
long, with 2,900 individual reefs and 1,050 islands. In
total area, it was larger than the United Kingdom, and
it contained more biodiversity than all of Europe combined. . . . The Great Barrier Reef was predeceased by
the South Pacific's Coral Triangle, the Florida Reef off
the Florida Keys, and most other coral reefs on earth. It
is survived by the remnants of the Belize Barrier Reef
and some deepwater corals.

Scientific journals now publish papers on "ecological
grief" as an occupational hazard of coral research, the "emotional vulnerability of scientists who work at the forefront of
an ecological crisis," and the importance of "recognizing how
ecosystem decline and climate-related events can affect mental
health." Suggested coping mechanisms include taking on side
projects researching mollusks and other species which are
more resilient to rising ocean temperatures than corals.[23]

One research project cited as a "reason for hope" by
National Public Radio and other media outlets is an effort to
replace dead coral with artificial coral made using iPhone
images of real coral and a 3D printer. The fake coral is printed
using either polyester, or cornstarch, or cornstarch combined
with stainless steel powder. The idea is to provide a sort of
emergency housing for reef-dwelling species when the coral
which formed their habitat die.[24]

"OK, all you fish have no homes, but we don't want you
to move away because we still want this economy to work,"
explained lead researcher Danielle Dixson. "So we're going
to put these temporary FEMA trailers up. You live in these.
While you live in these, actual, real corals will settle on them,
they'll get covered and eventually, they'll be a reef again. That's
the idea, at least."

Are Two Dimensions Enough?
The Networked Screen and the Human Imagination

The following essay was written in 2006 but never published until now.

In 2006, Facebook membership was restricted to students on a few US campuses. Twitter, Instagram, and Snapchat did not exist. YouTube had only just launched.

You couldn't "like" something on Facebook or any other app. "Social media" was not a generally understood term. Much less was it understood that it would soon eclipse all other media. And even less that it would soon be held in large part responsible for political catastrophes ranging from the election of Donald Trump in the US to the genocide of the Rohingya people in Myanmar. No one anticipated that Silicon Valley executives would soon express guilt for "ripping apart the social fabric of how society works."[1]

"Selfie" was not a word many had heard of or understood.

Wired magazine, the editors of which had proclaimed them-selves the in-the-know fortune tellers of the digital revolution, were predicting that computers would soon lead to an economy of permanent growth, free university education for all, and "a more rational, less dogmatic approach to politics."[2]

Much of the discussion about the cultural impact of digital technology centered on speculation concerning the transforma-tional potential of "immersive virtual reality" technology, the arrival of which was thought to be just around the corner. I was not convinced. Instead, I argued that flat, 2D, networked screens were already transforming culture in ways much deeper than was commonly understood. I suggested that the secret of their power was how different from the real world they were, not how closely they mimicked it. And that while this technology frequently entered the world as the second-best option that would do in a pinch when the "real" thing was unavailable, the screen-based activity quickly became the benchmark against which the shortcomings of the real world were measured.

When I made this argument, there were no iPhones in the world. Networked screens were confined to desktop and laptop computers. No one in the world carried a networked two-dimensional screen in their pocket. Very few imagined they ever would.

Making predictions about technology and culture is a fool's errand. Changes happen so fast, with wildly unpredictable consequences. The "future-ologists" who demanded our attention fifteen years ago because they were certain their predictions were spectacularly accurate now demand our attention because, they claim, they have learned from how spectacularly wrong they were.

In contrast, the arguments presented below are holding up well. In fact, as networked screens moved from our desks to our pockets they have become all the more compelling. Perhaps even more surprising, my arguments remain nearly as novel today as when I wrote them.

The essay is published here without revision, along with a new postscript.

"Immersive virtual technology" is here, all around us. It has already profoundly changed our desires, our sense of what is real, and our sense of ourselves. And though the pace of technological change demands that we attempt to anticipate its future consequences, analyzing the psychological impact of nonexistent technologies (nonexistent in the sense of a large number of people using them or having access to them) is a highly speculative endeavor. At the very least, we should begin by examining the impact of present technologies, lest our speculation degenerate into outright fantasy. Anticipation of changes that may be on the horizon must be grounded in the changes that can be observed and studied now.

Such an endeavor does not require a pilgrimage to the MIT Media Lab for a privileged first glimpse of futuristic design work, nor must we extrapolate from the experiences of a few people in test studies of early prototypes. Our culture is already "immersed" in a web of pervasive computational objects that mediate our psychological and social lives, our sense of reality, and of ourselves. The objects I am referring to are the ubiquitous two-dimensional networked computer screens that are an ever-present fixture in our lives today—not as their designers promise they will soon be, but as they exist now.

As opposed to the ambiguous and overwrought notion of "virtual reality," let's call it "screen reality." It generally makes a rather modest entrance as an understudy to plain old reality— understudy in the sense of something that can be used in a pinch when the "real" thing is not available. But once intro-duced, the understudy wins the starring role, as people using the screen-based substitute come to *prefer* it to the off-screen world.

Psychiatry in Networked 2D

The most obvious places to see this transformation are in the fields of internet pornography and games, but let's start a little further from public view, with telemedicine, and specifically telepsychiatry. Like so many present-day technologies, tele-medicine got its start as a NASA initiative, in this case to tend to the psychiatric needs of astronauts thousands of miles from a psychotherapist's couch. It got a big push from the US mili-tary, which was interested in remote control medical care of battlefield casualties. In other words, telemedicine began as a substitute for a visit to a real doctor, a second-best option to be used only in situations where a flesh-and-blood doctor was unavailable.

From outer space, telemedicine moved to confined space: prisons. Arizona and Texas have been leaders in the field. Over 80 percent of specialty medical consultations in Arizona prisons now take place via the screen. The doctors and patients might be no further from each other than across town (or even in the same building), but the experience would not be different if the prisoners were on a space station orbiting above the earth's atmosphere. Doctors comfortably isolated from their incarcerated patients use electronic stethoscopes to listen to heart beats and remote-controlled otoscopes to look at eardrums. Before 1997, Arizona prisoners needing specialist medical care were manacled and transported in vans or buses with all the extreme security procedures typically employed by American prisons. Not surprisingly, most prisoners came to *prefer* to receive their health care by telemedicine.[3]

From outer space and confined space, telemedicine moved into internal space: psychiatry. At least eight states now pay for telepsychiatry under their Medicare programs. Six states, including California, require private insurers to reimburse patients for telepsychiatry.

As with telemedicine, telepsychiatry began as a second-best alternative for when person-to-person psychiatry was not possible. Dr. Sara Gibson is a pioneer in the field. Her practice is in Arizona's Apache County, which has just sixty-nine thousand residents, high levels of poverty, drug use, child abuse, and suicide rates, but no psychiatrists.[4] She has had an exclusively screen-based psychiatry practice for more than a decade.

Mike Kueneman is one Dr. Gibson's telepatients. He says he feels more comfortable in his relationship with Dr. Gibson than he does with most of the people he knows, even though his dealings with Dr. Gibson have been entirely confined to the screen. They have never met in person. "It's hard for me to trust any other doctor," he explained. Of course, it is difficult for Kueneman to visit any doctor in person—or anyone at all for that matter. He is placed in leg shackles and handcuffs, and accompanied by an Apache County sheriff's deputy, simply to see a psychiatrist on a screen.

But preference for psychiatrists on screens is hardly limited to patients in leg irons. Nancy Rowe, who manages the program in which Dr. Gibson works, says that given the option of using telepsychiatry, her clients almost always prefer their psychiatrist to be an image on a two-dimensional screen rather than a physically present, breathing mound of flesh and blood.[5]

"Some people don't want to have to deal with a real person," said another of Dr. Gibson's patients, a sixty-three-year-old woman with dementia and bipolar disorder.[6]

Dr. Gibson says the physical presence of a psychiatrist can overwhelm a therapy session with an avalanche of sensory data that can distract patient and doctor alike. "Initially we all said, 'Well, of course it would be better to be there in person.'" she explained. "But some people with trauma, or who have been abused, are actually *more* comfortable."

It is worth noting how utterly novel this mental construct is. The idea that someone might be overwhelmed by the sensory data that comes along with another person's presence is new to human experience. Long-distance communication has moved from runners, ponies, and carrier pigeons, through the Postal Service, telegraph, and telephone. Telephones have been in widespread use in the US for a hundred years, and Americans have spent millions of hours conversing on them without wondering whether anyone was sensorily overwhelmed by the physical presence of another person. Nor has the idea waited for wearable computers, relational artifacts, or experiments with futuristic technology in the MIT Media Lab or other academic high-tech outposts. The idea appeared at a very specific historical time and place: when large numbers of people acquired access to networked two-dimensional screens.

This question in turn begets another: if psychiatry is indeed more effective without the "sensory overload" that the physical presence of another person entails, will telepsychiatry move beyond circumstances in which office visits are not viable, and conquer more and more of the world of psychiatry as a whole? If the sensory presence of another person is indeed detrimental, why should *anyone* visit a psychiatrist's office?

Sex in Networked 2D

Perhaps psychiatry is a form of human interaction uniquely suited to the screen. It was never supposed to be about physical touch or smell. The couch itself was intended to be something of a physical barrier between doctor and patient, so the idea that physical presence might intrude on the psychiatric exchange predates telepsychiatry.

What would be antithetical to psychiatry in this regard? How about sex? Surely sex does not *suffer* from the physical presence of another but rather *requires* it. However, we find the same pattern in sex we found in telepsychiatry: as pornography migrated to the networked screen, what had previously been a second-best substitute, for use when "real" sex with another living person was unavailable, is becoming for many the *preferred* mode of sexual experience.

We know that humans have been making sexual images and telling sexual stories as far back as the historical record of human artifact extends. Sexual content has flooded every new type of media as soon as it was available. It was not long after the invention of the printing press that erotic books landed in the *Index Librorum Prohibitorum*. Materials considered obscene by the standards of the day were enough of a problem in the American colonies for the Massachusetts Bay Colony to pass a statute against them in 1711. In 1953, *Playboy* became the first mass circulation porn magazine, selling fifty thousand copies in its first two weeks and reaching a peak circulation of seven million in 1972. In 1980 around a thousand "adult" film theaters in the US were showing porn films, and by the early 1990s videocassette pornography was a $490 million dollar industry.[7] The market demand for pornography has been a significant impetuous in the development and widespread adoption of both videocassette and internet technology.

While the steady growth in the porn market in the second half of the twentieth century is impressive, what happened to pornography with the arrival of the internet is of another order of magnitude. By 2003, the Internet Filter Review found that there were 260 million pornographic web pages, a 1,800 percent increase from just five years before. New porn pages were being added at a rate of 28 million per month. Twelve percent of all internet sites are devoted to porn. One-quarter of all daily search engine requests, or 68 million, were for pornographic material. Peer-to-peer downloads of pornographic material were estimated at 1.5 billion per month.[8] The internet traffic measuring service comScore estimated that 70 percent of eighteen-to-twenty-four-year-old men in the US visit adult sites each month, while the Hitwise measuring service found that the percentage of men in older age brackets who accessed internet porn was *even higher.*[9] If the internet really does represent the beginning of a "worldwide mind," then we know what it's thinking.[10]

It is not surprising that internet traffic measuring services only survey men when it comes to internet pornography. A 2003 study in Australia found that men were ten times more likely than women to have visited a porn site.[11] My use of gender in the discussion that follows reflects the fact that while use of internet porn by women is rising fast, it is still dwarfed by the use of internet porn by men.

The quantity of porn available at any given time and from any given computer with an internet connection is essentially infinite—no one could look through *all* those 260 million web pages (which by the time you read these words will total millions more). Contrast this with my hometown in Colorado in the 1970s, where there was one squalid porn movie theater, and maybe five or six porn magazines available in just a few stores. Those magazines were pretty much it for men in my town, and they had to wait a whole month before they could get a new supply of images and text. With internet porn there is no wait; more is always available, infinitely more.

This increase is so extreme that the quantitative change alone is having a qualitative impact on the way we fantasize about and engage in sexual activity. This can be seen most immediately in how we view our own bodies. Take, for example, the latest fashion in cosmetic surgery: vaginoplasty (surgical tightening of the vaginal muscles) and labiaplasty (reduction of the labia minora.) "There's remarkably amazing patient interest in this," says Dr. V. Leroy Young, chairman of the emerging trends task force for the American Society of Plastic Surgeons. Why? "Now women see porn," said Dr. Gary J. Alter, a plastic surgeon and urologist with offices in Beverly Hills and Manhattan who has come up with his own "labia contouring" technique. "Now they're more aware of appearance."

Dr. Bernard H. Stern, a gynecologist in Fort Lauderdale, began to focus exclusively on genital cosmetic surgery in 2000 and by 2004 was doing four to five surgeries a day on patients who come from all over the United States and abroad. "The women feel undesirable or unpretty," Dr. Stern explained.

One such woman was a thirty-nine-year-old yoga instructor who had a labiaplasty in Beverly Hills in 2000. The woman said she was "asymmetrical"—part of her inner vaginal lips extended about half an inch beyond the outer labia. "*The only women I could compare myself to were women in pornographic movies*," the woman explained. "They were tiny and dainty and symmetrical. Nobody looked like me."[12]

The increase in the quantity of porn and its ubiquity in our daily lives is not nearly the whole story. When contrasted with printed or videotaped porn, internet porn is not only more but different. The differences are the same as those that distinguish all information stored digitally and accessible over the web from information that is not, but these differences acquire a specific importance when the subject matter is sex.

Like everything else on the internet, internet porn can be accessed without going to a public place. The significance of this can be instantly grasped by any man who can remember trying to discretely exit a drug store, bookstore, convenience store, or smut shop with a pornographic magazine. Internet porn is available anywhere there is a screen—home, work, and now with the latest cell phones and video iPods—essentially anywhere. Digital porn can be easily concealed from nosy spouses. Truly massive collections of pornographic images, which in magazines and books would have spilled out of the closet and required whole rooms of file cabinets, can now be discretely saved in a locked folder on a laptop. One man interviewed for this essay had over 19 gigabytes of pornography stored on his computer. This number does not easily translate into a certain number of movie clips or still photos, as the size of each file will vary according to clip length and image resolution. But his 19 gigabytes certainly contains thousands of individual files. Compared with his online peers who share his Yahoo groups, he does not think his collection is particularly large.[13]

Another man explained that while he used to have to go searching through the web for pornography, he is now electronically linked to so many porn aficionados eager to share their collections that he gets well over a thousand new porn files a week emailed to him without even asking. For this man, the internet has become a sort of spigot with a high-pressure flow of pornography.[14]

It is common for internet porn collectors to have thousands more files than they could ever actually view. Some music collectors found themselves in a similar situation with music files in the early days of Napster and other peer-to-peer file-sharing sites, but intellectual property rights enforcement has restricted the free flow of music files. Porn companies themselves operate at the edge of the law and have been much less successful than record companies in enforcing their copyrights. And porn piracy has been supplemented by huge numbers of porn amateurs who photograph and post images of themselves having sex simply because they want to.

One result of the internet porn tidal wave is an explosion of sexual subcultures. Among the millions of porn websites are those that cater to every conceivable preference or kink. Consider a fetish so rare that only one in one hundred thousand people find it exciting. The potential market for porn related to this fetish could never have sustained a magazine, much less secured such a magazine shelf space in a town of, say, half a million inhabitants. On the average there would only be five people in town interested in the fetish, and the chances of those five actually finding the magazine on the shelf in whatever obscure shop would stock it would be highly unlikely. Even if they did, only five sales would result. But a website catering to this same fetish would find a potential market of three thousand among the three hundred million Americans, and of course it would be available to anyone in the world with an internet connection. Once up and running, the site can be stumbled upon by internet surfers meandering through the internet's pornography maze, people who might have some curiosity about the fetish but would never have gone to the extraordinary lengths required to explore this curiosity before.

The highest-profile sexual subculture in the US is the queer one, and there is indeed a marked decrease in homophobia among young people growing up with internet access. The same breakdown of taboo can be seen in the rapidly expanding market for sex toys, particularly among women. This growing market is benefiting both from the fact that sex toys, like images, can now be ordered privately without going to any public place, and also from the fact that pornographic images of people using sex toys are ubiquitous when one logs on to the internet.[15]

Images downloaded from the web, like all information that is stored digitally, are readily ordered and reordered, sorted and sequenced. Collections can be pored over in the most meticulous detail, sorted by body part, race, age, gender, activity, fetish, camera angle, body weight of model, and more. This sorting and sequencing may sound trivial, but I do not think it is. Creating and recognizing patterns is immensely satisfying to the human brain. Philosophers from Aristotle down to the present day have understood it as the basis for our appreciation of art. Interviews with collectors of porn images reveal an intense preoccupation with sorting and sequencing. Some say they spend more time sorting and sequencing than fantasizing or masturbating.[16] The combination of the human sex drive and the human drive to sort and sequence makes internet porn a powerful and unprecedented phenomenon.

There is no survey data to back up the following claim, but I have no doubt that if it were possible to add up the total number of male orgasms per year in the US, the data would show that sex between humans and machines has become more prevalent in our culture than sex between humans and humans. Do the math. More than 70 percent of men visit porn sites *at least* once per month. The whole point is to use this material for masturbation. At home, at the office, and now anywhere one can carry a cell phone or iPod. If one subscribes to the Catholic notion that sex in which semen is ejaculated into anything that is not a vagina for any purpose but procreation is sinful, then the internet is the devil's handiwork indeed.

A typical response to this claim has been, "But wait, does a man masturbating in front of a computer really constitute sex between a human and a machine?" My answer is simply this: what is in the room? Not one person touching another person. What is really there is a man touching himself with one hand and a machine with another. Given our existing cultural notions of machines, this still may not seem like sex between a person and a machine, but it is just these assumptions we need to question if we are to notice the cultural changes that are taking place right under our noses.

The next question is generally, "But would you also say that a man masturbating with a *Playboy* constitutes sex between a human and a magazine?" My reply is: who cares? Porn magazines were never available in infinite amount and variation, at any time and place, with endlessly sortable and sequence-able content. In short, they were not "interactive." Not "interactive" as in future technologies we might imagine but have yet to create, but "interactive" in the rather plain but actually existing sense in which I am speaking: image as information that can be cut and pasted, re-sequenced and re-sorted, on a flat 2D screen. It may be that this present-day notion of interactivity rapidly advances into the realms that futurologists predict, or it may not.[17] In any case it is changing our lives right now.

The ubiquity of the screen in American sexual culture is creating situations in which the experience of sex is a sort of layer cake, with layers of physical presence sandwiched between layers of screen. The website MilitaryClassified.com, for example, features video clips of a man, who is in the armed forces and allegedly "straight," sitting in a chair and watching straight pornography on a screen which is off camera. As he watches, a "gay" man fellates him. This whole scene is then recorded for the pleasure of the viewer. The layering here is as follows: at the far end is a screen showing porn. Layer two is a man watching that screen. Remarkably, he holds a remote control in his hand with which he can fast forward, reverse, or change tracks as his desires dictate, so there is touch between these first two layers of machine and human. Layer three is another man fellating the first man, so there is touch between these two human layers, even as the first man is touching the machine layer through the remote. The two human layers are being recorded on camera, so layer four is another layer of machine. The man in layer three is also the camera operator, and he continually breaks off his fellating to disappear behind the camera to make adjustments to the zoom lens to reframe the image being recorded, so there is another element of touch between human layer three and machine layer four. Finally, another person sits at home watching the whole thing via their internet browser, making a human layer five. More than likely, this human is clicking on his keyboard and mouse just like the man in layer two is doing with the remote, making for one last touching of human/machine between layer four and five. There is obviously no precedent for sexual activity of this sort.

I do not mean to read too much into the element of touch between human and machine, but neither should it be dismissed. One man interviewed for this essay had ruined six or seven computer keyboards "with just lube and DNA."[18]

The Understudy Steals the Starring Role

In sex just as in telepsychiatry, screen-based activity which began as a second-best substitute for sex with a real person is becoming, for an increasing number of people, the *preferred* sexual activity. As I mentioned before, it is impossible to state this trend in terms of quantitative data. The data does not exist, there would be no funding to acquire it, and most people are extremely reluctant to discuss with others the very private details of their sexual relationships with machines (another good indicator of the importance of the human-machine sexual relationship). But anecdotal evidence is plentiful.

There are numerous websites exclusively devoted to arranging immediate sexual activity between people who visit the sites, known in internet vernacular as "hookups." These sites are used mostly, but not exclusively, by men seeking sexual contact with other men. Many of the postings on these sites invite men to come to the advertiser's house or apartment and watch pornography. The men may or may not engage in other sexual activities as well, but the porn-watching is the entrée, if not the central activity. Postings on these sites can be extremely sexually explicit, disturbingly so to someone not acclimated to the subculture. For this reason, I have put some of these postings in the endnote at the end of this paragraph for those who are interested. If such things might offend your sensibilities, just take my word for it.[19]

The place of "real" sex and pornography has flipped for these men. Pornography was previously something that one looked at while fantasizing about being with another person. These men want to be with another person so they can fantasize about being in pornography. Pornography has moved from being the substitute to being the ideal.

This shift can be seen in even starker relief in anecdotes related to me by an assistant director at a major producer of gay internet porn.[20] He told of how, in just the few weeks prior to our interview, he had seen the following situation on more than one occasion: two men were having sex on camera during a porn shoot. One ejaculates. The camera stops and the second man masturbates, the idea being that when he is just on the verge of ejaculating, the cameras will roll again and the scene will be edited to make it appear as if the orgasms were nearly simultaneous. But the man masturbating takes a long time. Try as he might, he just can't seem to do it. Now, it is important to picture the scene here. The man masturbating is sexually attracted to other men. He has just been having intercourse with the man lying before him, and this other man has just ejaculated. This second man is there because he has been selected among many, many other men for his physical beauty. He's got "porn star looks." But the first man, the one who is masturbating, cannot reach orgasm. The film crew is getting impatient, since time is money on a film set. Finally, the crew brings a laptop computer and balances it on the body of the man who has already ejaculated. The laptop screen is showing gay porn. Watching the images on the screen, the man masturbating quickly reaches orgasm. Just before he ejaculates, the laptop is removed and the camera rolls.

I cannot imagine a more vivid illustration of the point. A gay man is confronted with the actual flesh and blood and sweat presence of another man who has spent hours at the gym sculpting his body into precisely the type that gay culture embraces as ideal (which is itself largely the product of pornography). But in order for the first man to reach orgasm, he needs to look at images on screens. What he sees on the screen is essentially the same thing that is lying before him in "real" life, except flattened down to a small image on a two-dimensional screen. The process of reducing the sound and sight and smell and touch of the "real" world to an image on a screen *increases* its erotic appeal.

So why have sex with another person at all? For the exploding number of "amateur" pornographers who are filling the internet with their millions of websites, the answer seems to be to create images to watch on the screen.

Who Passes the Ball?
Since internet pornography is generally viewed in private and is generally considered off-limits to serious research, and tele-psychiatry involves a relatively small number of users, computer gaming is the area in which the "immersive" nature of the human connection with the networked two-dimensional screen has been most commented on. I chose to discuss tele-psychiatry and pornography first, to establish that the close relation between human and machine as mediated by the net-worked screen is not unique to gaming, and more specifically, that "game play" is not a necessary ingredient of the human attraction to screens.

In the same vein, let's begin the discussion of games with Second Life, the computer game with no "game." Second Life is not a game but a "virtual world," a massively multiplayer web-based environment in which users create avatars which represent them, and then manipulate their avatar to interact with other avatars representing other users who are logged on at the same time. Second Lifers make virtual clothes, take virtual trips, build virtual houses, shops, nightclubs, casinos, and so forth—and do all sorts of things which bear little resemblance to other computer games. There is no competition. Users don't fight or shoot each other, there are no points, no winners or losers. Even the game's name speaks to the topic at hand, for the "second life" the product offers its users is represented in its entirety on a two-dimensional screen.

This screen-based world is vast: as of December 2006 there were close to three million "residents" from around the (real) world.[21] If you have not "visited" Second Life and are interested in the future of human-machine relationships, I highly recommend you give it a try, as it is extremely difficult to convey a sense of the experience in words. Second Life evokes the same sort of devotion, or fanaticism, or addiction, as other games that keep the user's adrenaline on high and their itchy fingers on the game pad trigger, but here there is no adrenaline and no trigger. When you visit the first time, you will be welcomed and oriented by an avatar who will show you around. This avatar is not automated, nor is it controlled by an employee of Linden Lab, the corporate entity behind Second Life. The welcoming committee consists of avatars controlled by other "residents" who volunteer to spend their free time orienting newcomers to their virtual world. When I first logged on, my greeter was an avatar controlled by a woman in Scandinavia, who gave me some general instructions, helped me outfit my avatar in some hipper virtual clothes, then invited me to her virtual home for some virtual nude dancing.

Nude dancing in Second Life involves staring at a two-dimensional screen on which a tiny icon, who represents you, makes clumsy animated motions that vaguely resemble dancing. Depending on the quality of your computer's graphics card and the speed of your internet connection, the dancing motion may appear to be a sort of jerky stop-and-start affair, or more likely it may clunk along from one frozen frame to another. Yet a California woman whose avatar is known as SweetBrown1 Mfume told a reporter she and her friends spend much of their free time in Second Life because, she said, "We can dance, hug and kiss, all across the US."[22]

Says another resident, "In real life, I love to shop. Here I get the same satisfaction, but it's more fun because you can pick the colors and it will always fit."

"Unless you're concerned with taste and smell, Second Life provides an . . . incredible tourist destination," explained Philip Rosedale, the founder and chief executive of Linden Lab. On another occasion Rosedale went further: "Our goal with Second Life is to make it better than real life in a lot of ways."[23]

It is easy to present these statements as ridiculous, but that is not my intention. Second Life is a powerful and creative endeavor in many ways. Everything in Second Life—the avatars, the real estate, the clubs, shopping malls, bars, and casinos—is created by its users. Linden Lab provides only the virtual space and the virtual tools. This is a new kind of software and a new sort of cyberspace that should be—and has been—analyzed from many angles. My interest here is presenting Second Life as a particularly striking example of how compelling networked two-dimensional screens have become; how more and more people, when confronted with the screen, are not much "concerned with taste and smell" or, for that matter, sound and touch.

Computer games based on real-world sports are yet another area where one can see a screen-based version of a real-world activity moving up from second-best to the preferred way to experience sports, both as participant and as spectator. This development is of grave concern to the National Basketball Association commissioner David Stern. "I was on a panel recently where someone asked me what my worst fear was. It was that as video games got so graphically close to per-fection, and you could create your own players—their hairdos, their shoes—that there might be a battle between seeing games in person or on television and seeing it play out on a video game."[24]

Stern is stepping up to the plate a bit late, as this battle is already engaged, and the screen-based games are winning. Since 2000, television broadcast ratings for almost all major sports have fallen among male viewers by some 34 percent, while sales of sports-themed computer games have risen by the same margin. For Joshua Alvarado, a sixteen-year-old who spends upward of six hours a day playing computer games, the battle has already been decided. "I love sports," he said. "But why would you rather just watch it on TV when the video game lets you control it?"[25]

The fact that the TV screen shows images of real people sweating and shouting and running and jumping in a real place, that figures on the screen dunking the ball are showing the accomplishments of real people who have spent countless hours training and practicing and coaxing the last ounce of effort out of real bodies, bodies very similar to Joshua's—none of this is as compelling to Joshua as the interaction he can have with the avatar in the computer game.

Joshua's preferences are shared by fourteen-year-old Albert Arce, who is a fan of the Los Angeles Lakers. "I like Kobe [Bryant], OK?," said Albert. "But I like to play him because I can make him pass to the other guys. When I see him on TV, it's like he doesn't know how to pass."[26]

Networked 2D for Kids

The age market for the games discussed thus far begins in the early teens, but our experiences prior to this age shape us in ways that stay with us for the rest of our lives. It is for this reason that Sherry Turkle, whose work lies at the intersection of digital technology and psychoanalysis, has focused much of her research on this age group, and specifically on "relational artifacts":

> Most recently, a new kind of computational object has appeared on the scene. "Relational artifacts," such as robotic pets and digital creatures, are explicitly designed to have emotive, affect-laden connections with people.... In the case of the robotic doll and the affective computers, we are confronted with relational artifacts that demand that human users attend to the psychology of a machine.... There is every indication that the future of computational technology will include relational artifacts that have feelings, life cycles, moods; that reminisce and have a sense of humor—that say they love us and expect us to love them back. What will it mean to a person when their primary daily companion is a robotic dog? ... In order to study these questions I embarked on a research project that includes fieldwork in robotics laboratories, among children playing with virtual pets and digital dolls, and among the elderly to whom robotic companions are starting to be aggressively marketed.[27]

The best-known "relational artifact" of the sort Turkle describes is the Hasbro Furby, a small furry toy resembling a stuffed animal that communicates with each other Furbies via infrared ports. Furbies speak with an electronic voice, and though they start out speaking "Furbish," over time they use Furbish less and English more. There was a common misconception that they could "learn" words they would hear, but this was not the case. Motors move the Furby's eyes, ears, and mouth, and raise it slightly off the ground.[28] The Furby became a "must-have" toy following its launch in the holiday season of 1998, with demand driving its price from the retail list of $30 to over $100 if you could find one. Over the next year twenty-seven million Furbies sold. But then the fad quickly faded. In August 2005, Hasbro relaunched the Furby, now upgraded with voice-recognition and more complex facial movements, but it failed to catch on. Many aficionados preferred the previous, "downgraded" version. Furbies are now commonly used by art students who take them apart and use the parts for art projects.

About the same time Hasbro relaunched the Furby, a company called Ganz introduced the Webkinz. Webkinz are not "relational artifacts" but plain old stuffed animals. But each toy comes with an ID number which identifies it with a virtual counterpart on the Webkinz website, a sort of Second Life for the elementary school set. Owners are said to be able to "discover" their pets' online personas. ("I'll let you in on a secret," reads the profile of a cocker spaniel. "I love fish sticks, and I've always wanted a bunny clown.") Children can buy clothes for their pets using virtual money, [and] decorate their pets' virtual rooms. . . . Children can also train for the instant messaging marathons of their older siblings' worlds by sending preset phrases to their friends. They can even invite pets over to hang out—virtually, of course."[29]

Ganz has skipped the motors, communication ports, and voice recognition software of the Furby and linked an old-fashioned stuffed animal to a virtual world on a flat screen. The small Canadian company has spent no money on advertising, yet in just two years they have sold over a million Webkinz, and demand is growing exponentially. When children speak of their Webkinz, they sound very much like adults speaking of Second Life. As one girl explained, playing with the stuffed animal is fun, but she prefers playing with her pets on the networked 2D screen "because there's a lot more to do."[30]

Trading Reality for Control

For years one of my main activities was making music with computers. I remember when the first music software for personal computers became available in the 1980s. The music created with this technology played with an electronically precise meter (the time between beats), making it sound radically different from the rhythm of music made by humans. Most musicians working with computers at the time had the same response I did: "Well, this is sort of cool, but it sounds so *machinelike* no one will ever listen to it. To be really interesting, this technology is going to have to evolve to sound more human." And software engineers busied themselves trying to make computer music sound human. But before they could solve the problem, a new generation of kids had come up who *preferred* the machine-like quality of computer music. Music with an electronically precise, not humanly possible meter has now flooded the world.

The nexus of innovation at the meeting point of humans and their technology works from two directions: on the one hand, our present tastes and desires lead us to create new technologies. On the other hand, our tastes and desires adapt to the tools we have available. The quickening pace of technological change has garnered a lot of attention. Much less noticed has been the acceleration of change in our tastes and desires as we adapt ever more quickly to the changes in our media-drenched world. In the case of electronic music, as fast as the technology changed, our listening preferences changed even faster.

This dynamic extends far beyond music. Much like early computer musicians, computer design specialists have looked at early versions of screen-based activities such as sex and games, and thought, "Well, that's cool, but it's limited to a flat screen, and will only interest people so far. To really grab people, this technology needs to become more like what people think of as 'real' experience."

Engineers have been hard at work at just this task, designing computers that are wearable, affective, emotive, robotic, and to some extent intelligent. There are prototypes of devices that strap in the user's body and manipulate it in real space to correspond with the motions of the user's avatar in the virtual world presented in a helmet on the user's head. Over 150 different models of "virtual reality helmets" with 3D displays and surround sound audio have been marketed.[31] Meanwhile, a steady stream of souped-up screen products has become widely available. There are screens that respond when spoken to, screens that come with high fidelity sound systems, screens you can draw on with a pen, and screens that respond to the touch of a finger. You can present 3D images on a screen instead of 2D images. As of this writing, Apple has just announced a cell phone with a multitouch screen (you can touch it simultaneously with more than one finger). Some of these additions involve technology that has a very high gee-whiz factor. But judging by sales figures, not one of these devices has made the experience of interacting with a networked screen significantly more compelling. The one innovation to the networked screen that has been a hands-down success has been to simply shrink it down to where it can fit in your pocket: text messaging with cell phones has exploded in popularity, particularly among teenagers and young adults. Here again we see people choosing to bypass a technology that is readily available and closer to "reality" (transmission of the sound of a human voice) in favor of the decidedly less "realistic" alternative of communicating via text on a networked screen.

Just as with music, people's tastes and desires are changing much faster than this technology, acclimating to machines as they exist now, not as their designers promise they will soon be. Not only is electronic music preferable to human music, screen sex is becoming preferable to sex with a person, screen games preferable to sports, psychotherapy via a screen preferable to the "sensory overload" entailed by the physical presence of a therapist.

Taken together, these developments pose the question of whether responding to the physical presence of another human being as an "overwhelming sensory overload" is unique to trauma victims in psychotherapy, or is it becoming a generalized condition of human experience in a networked screen-based culture? Recall the Linden Lab executive's statement that "unless you're concerned with taste and smell, Second

Life provides an . . . incredible tourist destination." So, how concerned with taste and smell, and touch are we?

No one can touch or taste or smell the bodies they see in internet porn, but they get a degree of control of a sexual encounter that they could never get from another human. They can see exactly the image they want, when they want, and when they grow bored they can simply log off with no emotional obligation to their partners. Computer gamers can likewise play a game exactly how long they want, exactly how they want. When the screens are networked, they can view avatars or webcam images controlled by others in real time, and the experience becomes even more compelling. Much of the hype about the internet was that it would bring people together, but in fact much of the appeal is that it keeps us apart. When I send a text message, I know that the recipient cannot infer anything from my tone of voice or an awkward pause. If my sex partner can only see me on my webcam, I can reveal only the part of my body I want to and no more. Networked screens allow us to interact with others while maintaining a strict isolation of our personal physical space. *The appeal of screens is not that they emulate our customary lived experience, but that they interrupt it.* Rosedale says it's "like Burning Man, it's like taking a drug. Both of us are here, but we're much more comfortable with each other."[32]

Roy Brown spent fifteen years in prison in New York for a murder he did not commit. While in jail, he managed to conduct his own investigation from his cell that identified the actual murderer and won a court order for DNA testing, which proved his innocence and eventually secured his release. At the court hearing at which Brown was released, it was revealed that the detective who had originally investigated the murder suppressed evidence pointing to the real murderer as well as other evidence exculpating Brown. During his years behind bars, Mr. Brown became gravely ill with liver disease, so it was difficult for him to stand up when the judge issued a rather curt apology and released him. For the first time in fifteen years, Roy Brown stepped out into the world of fresh air, tasty food, sunrises and sunsets, music and dancing, and comfy beds to sleep in. He told the waiting crowd of relatives and reporters, "I can't wait to play Pac-Man."[33]

Your Last Linden Dollar

My focus here has been how existing technology is changing the human experience now. Having framed the essay within that constraint, I will hazard some brief comments on how the dynamics I have discussed *might* play out in the future.

The outlines of the possible future I wish to highlight can be detected in the seamless transition that has already occurred in sports on screens, from images of real athletes captured by cameras to computer generated avatars in virtual games. Anyone watching these avatars immediately sees that they are not images of real people. To someone of my generation, they look like "cartoons." The fact that they are "generated" in 3D rather than "drawn" in 2D computer graphics doesn't make them any more lifelike, it just makes them cartoons of a different sort. Yet people who enjoy watching sports on screens have been moving from watching sports on TV to playing sports-themed computer games in droves, a shift of some 30 percent of the total market in just six years.

What would prevent a similar shift in pornography, from real actors captured by camera to computer-generated avatars? All the factors that might inhibit such a shift, beginning with the fact that the avatars would not look convincingly realistic, would not represent real people, would not have human intelligence, and so on, are equally present in the field of sports, yet the transition in sports is well underway. What would be the social consequences of such a change? Will gay men seek out the company of others to share fantasies of having computer-generated bodies? Will women want labiaplasties to make their vaginas look more like computer-generated vaginas? Will the pinnacle sexual experience by which people measure their own sex lives be scenes between virtual avatars?

And what of telepsychiatry? In the 1960s, Joseph Weizenbaum created the first software that could "converse" in English. Named ELIZA, the software imitated—or, as Weizenbaum insisted, *parodied*—a Rogerian psychotherapist. Weizenbaum took advantage of a limited range of psycho-therapy techniques to make a program that, while remarkably simple, could mimic therapists to a certain extent, as in this example:

> Men are all alike.
> IN WHAT WAY
> They are always bugging us about something or other.
> CAN YOU THINK OF A SPECIFIC EXAMPLE?
> Well, my boyfriend made me come here.
> YOUR BOYFRIEND MADE YOU COME HERE.
> He says I'm depressed much of the time.
> I AM SORRY TO HEAR YOU ARE DEPRESSED.

It was simple to feed ELIZA a line that triggered a nonsensical response and revealed its limited capabilities, and Weizenbaum only intended to use the software to demonstrate principles of computer programming for visitors to MIT. Weizenbaum was shocked to discover that ELIZA provoked a serious discussion among his colleagues as to whether further enhancements to the software could create a product that would be useful for psychotherapy. Weizenbaum was so disturbed that he left MIT and wrote a book, *Computer Power and Human Reason*, explaining exactly why he thought computer software could never be a useful therapist.[34] But Weizenbaum's original col-laborator on ELIZA, Kenneth Colby, took the opposite position:

> A human therapist can be viewed as an information
> processor and decision maker with a set of decision

rules which are closely linked to short-range and long-range goals.... He is guided in these decisions by rough empiric rules telling him what is appropriate to say and not to say in certain contexts. To incorporate these processes, to the degree possessed by a human therapist, in the program would be a considerable undertaking, but we are attempting to move in this direction.[35]

Colby wrote this in 1966, during the initial excitement about artificial intelligence, when some researchers—notably those at MIT—thought the development of computers with an intelligence equal to that of people was just around the corner. As it turned out, this forecast was wildly overoptimistic (or pessimistic, depending on one's point of view). As the push for artificial intelligence ran out of steam, the debate about computers as psychotherapists faded from relevance.

In recent years there has been a resurgence in interest in machine intelligence, fueled by fast computers and research in emergent systems and neural networks. The dream of machine intelligence that rivals or even surpasses that of humans is once again a lively topic. My intention here is not to address that debate but to sidestep it. The developments I have described suggest that computers may not need to become all that intelligent for people to turn to them for something like psychotherapy. If Dr. Gibson is correct that most patients prefer telepsychiatry over psychiatry in the physical presence of a psychiatrist—that a psychiatrist on the screen is preferable, just as a sex partner or athlete on the screen is preferable—why must that two-dimensional screen therapist be a representation of a real person? If patients feel more comfortable confiding to a therapist that does not "overload" their senses with the presence of another person, why would they not feel even more comfortable with a therapist who had no connection to another person at all?

The debate between Weizenbaum and Colby assumed that the answer to this question had to await the arrival of machines with human-like intelligence. But there was a similar assumption that the acceptance of computer music required machines that could play music like humans, yet the world is now awash in music made by machines that is played the way machines play music. There is a mass exodus underway of spectators away from screen representations of human sports, in which both sides feature athletes with human intelligence, to watching screen representations of humans playing games with computers of extremely limited, decidedly unhuman intelligence.

Ten years after the original fuss about ELIZA, Carl Sagan commented:

> No such computer program is adequate for psychiatric use today, but the same could be remarked about some human psychotherapists. In a period when more and more people in our society seem to be in need of psychiatric counseling. . . . I can imagine the development of a network of computer psychotherapeutic terminals, something like arrays of large telephone booths, in which, for a few dollars a session, we would be able to talk with an attentive, tested, and largely nondirective psychotherapist.[36]

The internet has put all those terminals Sagan imagined in our homes, and now cell phones have put them in our pockets. Sagan imagined these digital therapists as being fairly smart; I picture them being pretty stupid, yet will their users mind? I wouldn't bet my last Linden dollar on it.

Postscript, 2020

Fourteen years after the above essay, VR headsets remain awkward curiosities, and 3D "caves" once thought to be on the verge of taking off sit largely unused in a few universities and museums. No matter. The migration of human behavior from the human-human nexus to that of human-machine rushes onward.

Three of the top ten websites in the world are porn. The only websites getting more visits than the top porn site are Google, YouTube, Facebook, Baidu, Twitter, and Instagram. Not even Amazon can successfully compete.[37] Take just one website: Pornhub. Over the course of 2019, Pornhub had 42 billion visits (an increase of 8.5 billion total from just one year before), which averages out to 114 million visitors per day. That is the equivalent of the entire populations of Canada, Australia, Poland, and the Netherlands, from newborn babes to grandparents, all visiting in the same porn site in one day, every day of the year. Nearly 7 million new videos were uploaded in 2019. If you strung together all the porn videos uploaded to this one site in just the past year and started watching them, it would take 170 years to watch them all.[38] And Pornhub is only the *third* most popular porn site on the web.

The sexual identities claimed by humans have yet to catch up with the fact that most sexual encounters now occur between humans and machines. Is a "gay" man who spends more time having sex with machines than with men still "gay"? What if most of the porn he watches is "straight"? Does he require a new letter in the ever-expanding list of LGBTQIA . . . ? Is he a techsexual?

More than three quarters of all porn is now watched on smartphones, which didn't exist at the time of the essay. In other words, the large majority of porn is watched on tiny screens that are far less like the real world than the bigger screens on desktops and laptops but offer more control in their portability. One of the main arguments in the essay was that control, not realism, is at the core of the appeal of the networked 2D screen.

The growth of the computer game market has mirrored the porn market. In 2006, the global computer game market was estimated at US$31.6 billion.[39] In 2019 that figure was US$123 billion, collected from 2.5 billion gamers (one out of three living humans).[40] Well over half of that revenue came from games played on phones. Here again, the greater realism of larger screens was traded for portability and control.

Second Life has been eclipsed by much bigger fish in the game world, seeming positively quaint compared to the current offerings. Yet it continues to log 350,000 new user accounts per month, and Second Lifers paid $68 million for virtual in-game goods in 2017, a number that continues to grow each year.[41]

Game play consoles have become a major way consumers view porn. This market segment is important enough that engineers at Pornhub "worked tirelessly to ensure that Pornhub would be fully supported on the PS4 and Xbox One systems at their launch (along with continued support for the PS3 and Wii)." Compared to porn viewers on laptops or smartphones, gamers tend to spend more time, watch more videos, and have a lower bounce rate."[42]

More than just viewing porn on game consoles, the realms of porn and computer games have begun to merge in other ways. Computer game titles and characters are now among the most searched-for terms on porn sites. In other words, many users have switched from searching for "teen" or "anal" on their favorite porn site to searching for "Final Fantasy" or "World of Warcraft." The search will return long lists of animated porn videos using characters from the game, or porn shot with actual humans dressed as game characters. In fact, game-themed videos is almost the only corner of porn in which costumes, dialog, and plot can still consistently be found. You can sit for hours in front of a giant screen holding a gamepad and playing a game, then use the gamepad to jump from the animated game to animated porn featuring the same characters you were watching in the game. Then jump to humans having video sex in those same outfits, then back to the game—all without getting up from the couch.

One development I completely failed to anticipate was the emergence of porn abstinence as a rallying cry of white-supremacist street fighters. (Of course, I wrote the essay in 2006, two years before the election of Barack Obama, years before the emergence of white supremacism into mainstream politics.) Take the Proud Boys, who have been facing off against Antifa activists in street battles in several US cities. There are four levels of Proud Boy membership. First is to declare oneself a Proud Boy. Second is swearing off internet porn. The idea is that "defending the West against the people who want to shut it down" requires marshaling all available toxic masculinity for the fight, so no more shooting it all around the bedroom in porn-induced waste.[43] The politics is easy to ridicule, but the takeaway is that porn obsession is a fact of life that is haunting a generation.

At the end of the discussion of pornography in the original essay, I asked: "So why have sex with another person at all? For the exploding number of 'amateur' pornographers who are filling the internet with their millions of web pages, the answer seems to be to create images to watch on the screen."

Several years after writing that, I met a gay man in his twenties who told me he had been masturbating to internet porn since he was fourteen years old and had had sex with another person just twice. Both times were on camera, filmed by a porn company, and posted to the web. Here was the man I had hypothesized: a man who generally has sex only with machines; the only reason for him to have sex with another person is to generate more images to load into machines.

And several years after that, "for fans only" porn appeared, referring to sites which allow anyone to start making their own porn and charging a monthly fee to view it. The largest such site, OnlyFans, launched in 2016, had more than sixty thousand content creators (only some of whom were porn creators) and millions of registered users worldwide in just three years of operation. Two new posts per week seem to be what viewers expect for the subscription fee they pay to creators, and creators report that recording, editing, and posting sex videos twice a week leaves no time to have sex that is not performed on camera for the purpose of posting.[44] That adds up to thousands of people on just one platform who only have human-human sex for the purpose of generating images for use in human-machine sex. How quickly that first person I encountered for whom that was true multiplied into tens of thousands.

Completely new since my original essay is the emergence of eSports. Like other major sporting events, eSports competitions are held in sold-out stadiums and broadcast worldwide. But instead of watching people kick or bat or bounce a ball, eSports spectators watch people sitting in front of screens competing at computer games. Viewers at home look at screens showing stadiums full spectators staring at giant overhead screens, which show the contestants at center stage staring at their own screens.

Starting from essentially nothing in 2010, the eSports market hit $779 billion in 2017 and is projected to top $3 billion by 2025. That is still less than half of the global market for the NBA, but it is growing at nearly 20 percent per year. Some pro soccer teams have signed up whole squads of players in a number of different eSports. "The thinking is simple: digital gaming is where the next generation of fans will come from (often, a young person's first interaction with a professional football club is through the Fifa game), and so eSports are a vast reservoir of future income."[45]

Telepsychiatry has evolved on a similar scale, with its own implications for who we are becoming. In 2006, these practices were still known as telepsychiatry, with the reference being the television, or telephone, or for that matter the telegraph. Today these practices are known as eCounseling, or cyber therapy.

One eCounseling site, Talkspace, claims one million users. BetterHelp, a competing service, claims 740,000.[46] How many therapists do you have to choose from where you live? Ten? Fifty? One hundred? The eCounseling sites claim a total of 11,000 therapists. Your therapist can be in Bangor or Bangalore. You can continue your therapy no matter where in the world you travel. Therapy for the expat lifestyle.

And it's a bargain. The therapist pays no office rent and competes with 10,999 others offering the same services. "Feeling better starts with a single message," says Talkspace, after which you can "start improving your life today for as little as $65 a week."[47] And since "emotions can't be scheduled, with Unlimited Messaging Therapy™ you can message your personal licensed therapist exactly when you feel like it." Talkspace therapists respond once or twice a day.

A hundred years of research on why the therapist's couch should be placed just so, or why the therapist should sit to the side or the back or the front—scrap all of that. That discussion now concerns the relative merits of video, voice, and text message.

Every time we use our smartphone to communicate with someone, we make a decision about whether to use video, voice, or text. The winner most of the time is text. I do not see any reason why eCounseling should go any differently. Whether within family, among friends, or between therapist and client, the appeal of texting is not how closely it replicates the experience of in-person communication. The appeal is that it is so different.

But once we get to psychotherapy by texting, the inevitable next question is this: *why have a human therapist at all?*

The forty years between the creation of ELIZA in the 1960s and 2006 when my essay was written saw very little advance in the field of artificial intelligence. Not so the fifteen years since. The capabilities of AI have grown by leaps and bounds. We talk to our phones and home speaker systems now, using the same grammar and vocabulary we use to converse with each other. The machines are smart enough to answer back using the same. Computer-generated avatars and voices are still obviously computer generated, but the text messages sent to your screen by a person look exactly like those sent by a chatbot.

The first serious attempt to create a chatbot therapist began in 2017 with the launch of Woebot, a cognitive behavioral therapist in the form of a text-interface chatbot. More than fifty thousand people talked with Woebot during its first week on the job. As its creators noted, that is more than a human therapist could talk to in a lifetime. Two years later, humans were exchanging between one and two million messages a week with Woebot.

No user will be fooled into thinking that there is a real person behind Woebot. For starters, Woebot is available at all hours and will remember everything you say in every session. Nope, definitely not human. But those are pluses, not minuses. Many users report that they prefer talking to a bot over a human, because they don't feel judged.

And there you have it. The ability to judge assumes agency. People want a judgement-free therapist. We turn to a computer not because it has agency, but because it does not. Not because it is like a human, but because it is like a computer.

What could possibly go wrong?

The Woebot privacy policy states that the company "may collect information about you from [third-party applications] that you have made public via your privacy settings. This supplemental information allows us to verify information that you have provided to us."[48] So Woebot is going to know more about you than your human therapist was ever allowed to know. But there's even more: "If you participate in the special program, we will share the outcome of your participation in the program . . . with the program partner, which may include your employer, certification authorities, or other medical and academic partners who help conduct the study."

Great. I wonder how many Woebot users have read this part of the agreement.

Conversations with Woebot within Facebook Messenger are subject to the Facebook privacy policy. Facebook can see that you are talking to Woebot, and they can see the content of the conversations.

That's reassuring.

★

In 2019, a seventeen-year-old girl was scrolling through her Instagram feed and was confronted with a Talkspace ad. It is not surprising that the Instagram/Facebook ad algorithm had selected an eCounseling ad for this girl, as her older sister had committed suicide four months before, so her social media accounts may have been full of grief. The ad showed a chat screen between a Talkspace therapist and client. The client had typed, "Our family dinner was really overwhelming. I feel like I can't be my real self during the holidays."

The Talkspace chat screen shows face pics of the people chatting. In this ad, the face of the person who could not be her real self during the holidays was the face of the seventeen-year-old's girl's dead sister.[49]

Instagram screen grab. The picture of the deceased sister's face is blurred out for privacy.

Technics Turntables
and Civilization
Many Would Die to See Her Live

We talk about live music all the time. We say, "I went to a show of live music last night." Or "I love live music." Or "I miss live music." But what do we mean? Do we mean anything? Or have the words become a sort of empty chair, where a meaning sat before wandering off when we weren't looking?

When people first started using the term "live music," they meant music that was not recorded.

I met a young man who was excited about the new synthesizer he had recently purchased. The device had a "demo" button, and he showed me how he could push it and completely serviceable electronic dance music would pour forth. I ran into him a few months later and asked him how the music was going.

"Great. I started out with that 'demo' preset and developed it into some stuff I liked so much I got it pressed into vinyl. The other night I went to a club where a DJ was playing who I had given my record to, and it was so exciting to hear my music played live."

For some time now, when I hear someone say "live music," I ask what they mean.

If they have difficulty considering the question in the abstract, I ask if they think DJs play live music? Most say yes.

OK, I reply, what if I walk on stage with a laptop and press play? Is that live music? Most people say no.

OK, I reply, what if I press play and turn a knob?

What if I press play and turn two knobs?

How many knobs before it's live?

Knob Syncing

"Live" or not, turning knobs on stage has become a fantastically lucrative activity for its celebrity practitioners. The ten best-paid DJs now pull in more than $25 million a year between them. Despite all the hand-wringing over how difficult it is to make a living playing music in the digital age, these are some of the best-paid musicians in history. And they routinely play in front of some of the largest crowds in the history of music.

They put on extraordinary performances. They dance energetically. They pump their fists in the air. They yell to the audience to pump their fists in the air. They jump up on the table. They jump down from the table. They shout out to the audience to "TAKE OUT YOUR PHONES" so that everyone can get their own video of the magic of the moment (and turning the inevitable cell phone use from a distraction from the ritual to an essential part of the ritual). Every now and then they turn knobs on the electronic gear in front of them. And sometimes, the knob they turn actually results in some sort of change in the music. But often the knob turning is just for show, as the music is played back from a prerecorded track.

Like lip syncing, but with a knob instead of your mouth. Knob syncing.

The curtain was pulled back on knob syncing, or at least torn a bit, by Joel Zimmerman. Better known as Deadmau5 (pronounced "dead mouse"), Zimmerman is one of the highest-paid DJs in the world, and thus one of the highest-paid musicians in history. He also tends to blurt out things that others would leave unsaid. In a 2012 *Rolling Stone* interview, he noted that one of his million-dollar-DJ friends "has two iPods and a mixer and he just plays tracks," while another "has a laptop and a MIDI recorder, and he's just playing his shit. . . . People are, thank God, smartening up about who does what—but there's still button-pushers getting paid half a million."[1]

In response to all the indignant blowback he got from offended DJs, Zimmerman doubled down with a blog post:

> It's no secret. when it comes to "live" performance of EDM. . . . It's not about performance art, its not about talent either (really its not). In fact, let me do you and the rest of the EDM world button pushers who fuckin hate me for telling you how it is, a favor and let you all know how it is.
>
> I think given about 1 hour of instruction, anyone with general knowledge of music tech . . . could DO what im doing at a deadmau5 concert. . . . honestly. who gives a fuck? i dont have any shame in admitting that for "unhooked" sets, i just roll up with a laptop and a midi controller and "select" tracks n hit a spacebar. . . . im not going to let it go thinking that people assume theres a guy on a laptop up there producing new original tracks on the fly. because none of the "top dj's in the world" to my knowledge have. myself included. . . . But to stand up and say youre doing something special outside of a studio environment, when youre not, just plain fuckin annoys me.[2]

What sort of facial gestures are appropriate while knob syncing in front of hundreds of thousands of people? What might be the authentic expression of someone turning a knob for "real"? The expression that a skilled knob sync artist would mimic? The problem goes even beyond that, however, because the music is not a recording of a performance. It is a spliced-together collection of machine processes. So the performer is not mimicking a human performance but a machine process. Zimmerman has solved this riddle by performing exclusively in a Mickey Mouse helmet which completely hides his face. Though it is probably best to not refer to it as a Mickey Mouse helmet, because we don't want to stumble into the trademark war that erupted between Disney and Zimmerman.

(Zimmerman also tried to trademark the name of his cat, a name already trademarked by a woman selling cat merchandise whom Zimmerman vowed "to litigate out of existence."[3] Which is even weirder considering that Zimmerman shamelessly stole the helmet-over-the-head schtick from the electronic music duo Daft Punk.)

The helmet-over-the-head solution to the problem of facial authenticity while knob syncing proved effective enough that many other mega-paid DJs have followed suit, including Marshmello (an upside-down bucket with a happy face), Claptone (a golden mask with a bird beak), Cassette (cassette-tape inspired), Malaa (a black ski mask), and so on.

Zimmerman was asked if he could see out of his mouse helmet.[4]

No, and that's what fucks me up. . . . One time I thought I killed a kid. . . . I hear something in my IEM (In-Ear Monitors) that was all broken up, and it sounded really urgent: Some kid broke onto the stage and started running out the back. He just [wanted to get up there], but I don't know that because I'm wearing a mouse head, and I can't see shit. Even worse, I'm looking out of a camera down here [motions to his chest] and I've already got a complex. . . . I did a backwards kick, and this kid went flying off the ladder and boom, right on his back. I turned around, and I took the helmet off and

I saw this kid just fucking laying there. I thought I was done, that I broke this kid's neck.

We can debate the ontological status of the rest of Zimmerman's performance, but the backwards kick was certainly live. Fortunately the kid was not dead.

So while Zimmerman is on stage, his connection with the world outside his mouse helmet comes from speakers placed in his ears and a visual feed from a camera mounted on his chest.

Writing in the earliest days of sound movies in the 1930s, the German art critic Walter Benjamin described the sight of a living actor surrounded by the technology on a movie set as "an orchid in the land of technology." Ha! He had no idea.

Knob syncing is easier than lip syncing. Everyone knows how a mouth moves when a singing voice comes out of it, so effective lip syncing takes practice. If you open your mouth at the wrong time, your mistake is obvious. Top-of-the-line drag queens practice this for hours.

The neat thing about knob syncing is that no one has any idea what a well-synced knob performance would look like, or even when a knob turn should occur, since no one knows exactly when knobs were turned during the production of the track in the first place. Or if any knobs were turned at all. Maybe the track was made entirely by mousing or by typing on a computer keyboard.

So knob syncing is a piece of cake. You can't get it wrong. You can do anything! You can turn any knob anytime. You can jump around, pump your fists in the air, scream and yell, and turn a random knob whenever you can fit that move in. You don't even need to be able to see out of your helmet. Artistic freedom!

Note that the question here is not whether turning a knob to change the sound of the playback of recorded music constitutes live performance. The question is whether turning a knob that you *pretend* changes the sound of the playback of recorded music constitutes live performance.

But really, who cares?

Let's say you are packed among the four hundred thousand people who managed to score a ticket to the mammoth Tomorrowland festival.[5] You are confronted by a video screen two kilometers long, several thousand lights, a sound system of thousands of watts, and a massive pyrotechnics display, all of which took four weeks and 1,500 people to set up. Maybe you got the cheapest ticket (just over $500) and are camping in a little tent in the outer burbs of the festival ground. But then again, maybe you sprang for one of the "mansions" on the festival grounds that run $50,000 for a group of twelve, and then had a meal at the "secret resto" hidden inside the main stage for another $25,000.[6] And you flew to Belgium on one of the many commercial jetliners chartered by the festival.

Or maybe you are at a more modest festival, like Electric Daisy Carnival Las Vegas, with its three hundred square meters of video screens and two thousand light fixtures, which took two weeks just to build the supporting structures for, followed by a week of light setup, another week for video setup, followed by days of programming.[7]

Anyway, here you are, packed in, shirtless, high, drenched in sweat, and there is a guy at the center of it all at what appears to be an altar. And, yes, it is nearly always a guy, and for that matter nearly always a white guy.[8] (That is a whole other story, and rarely spoken of: how the black DJ innovators of the 1970s were replaced by the mega-paid white DJs of today.) Anyway, he's up there, surrounded by every sort of fire and light present-day technology can muster, wearing the vestments of the true nonbeliever. He is fiddling with a knob or two on some electronic equipment which is about as opaque to you as a Catholic mass said in Latin.

Why would you care whether the knob turning is "real"?

For that matter, *why would you care that there is a DJ at all*?

Why aren't two kilometers of video images, thousands of lights, pyrotechnics, and several hundred state-of-the-art loudspeakers enough? Why the need for the guy in the mouse helmet, pumping his fists in the air and knob syncing?

Before Live Music

Until the end of the nineteenth century, it was obvious why you wanted a living human involved in a musical performance. The energy to power a sound wave through the air had to come from somewhere, and that somewhere was human bodies. Someone had to hit something, or blow on something, or pluck a string, or scrape some horsehair across a cat gut, or force air from their lungs through their larynx, and so on.[9]

There was no "live music." There was only music. "Live music" emerged as the counterpart to recorded music, which first appeared in the late nineteenth century and took off in the early years of the twentieth. You could listen to recorded music from a disc or music made by human bodies. All music which was not recorded was "live."

In the latter half of the nineteenth century, most of the music that was heard in the US was played on pianos, and more than twenty-five thousand new pianos were sold each year. Hundreds of American piano manufacturers employed thousands of workers.[10] By 1887 over five hundred thousand youths were studying piano, and by the end of the century seven out of ten public school kids were being taught to read music.[11] There were more pianos and organs in the country than bathtubs.[12] Pianos were to the nineteenth-century home what record players, radios, and then televisions would become in the twentieth century: the most important piece of furniture, around which families would gather in the evening. Having a piano placed prominently in the home, and having a child who could play it, became a signifier of middle-class stature and taste. A half century later, 1960s television star Dinah Shore opened a show dedicated to the piano noting, "Of course practically everybody in America at one time or another has tried or has played the piano. Most parties begin or wind up there, and I'll bet the first time you fell in love there was a piano playing nearby."[13]

Providing sheet music for hundreds of thousands of new pianists was big business. Most of it was published out of one block in Manhattan known as Tin Pan Alley. Sheet music had "hit songs" in the same way the record industry would later. One such hit, "After the Ball," sold two million pieces of sheet music in 1892 alone.[14] Factoring in population growth, "After the Ball" sold about as well as hit records by Lady Gaga, Justin Bieber, and Drake sold in 2015.[15] Companies even issued sheet music to advertise early consumer products like Bromo-Seltzer, foreshadowing the advertising "jingle" that would fill the airwaves once radio and television came along.[16]

The first record players were introduced into this world of millions of Americans playing hit songs on home pianos, reading from sheet music. Sheet music retained its dominance for years. The record of a hit song was viewed as a promotional device for the song's sheet music and was usually released only after sheet music sales began falling. Sheet music continued to outsell records of the same hit songs, and usually by a wide margin, until 1921.[17]

Lip Syncing

Lip syncing arrived with live music, as a sort of midwife. That's right: lip syncing was present right at the birth of live music and its twin, recorded music. It arrived with great fanfare, the central gimmick in a massive promotional campaign run by the Thomas A. Edison Company to sell early record players.

The first lip syncing took the form of a concert hall recitals in which "an artist's living performance" would perform a duet with "its RE-CREATION."[18] Audiences were challenged to distinguish between the two. Due to the clever way the spectacle was presented, this was a more difficult task that one might think. The vocalist was selected by the company for her ability to mimic the sound of her recorded voice. The record player always started first. There was a lot of surface noise on early records, and all that crackle and pop would have been a dead giveaway that the record had begun playing if the singer had started first. Once the record was playing and the recorded voice was singing, the living vocalist would start singing along with the recording, then drop out, then drop back in again.

For extra effect, at some point she (it was nearly always a she—early record players, like the latest gadgets today, were seen as technologically sexy) would stop singing but continue mouthing the words. The audience would not notice until the record stopped and the singer was left alone moving her mouth in silence. Or the converse: the vocalist would sing along with the record, the stage lights would go off as the music continued, and then the lights would come back on to reveal only the record player on the stage. A stunned audience would then realize they had not noticed when the singer had left.

There was a certain amount of hucksterism in these "tone tests," as they were eventually called by the Edison Company. But those early records, minus all the surface noise, did a surprisingly good job of accurately reproducing sound, just as the early forms of photography produced sharp images that compare favorably to today's cameras, though the images could only be black and white, and no copies of the original photograph could be produced.

Beginning in 1915, the Edison Company presented over four thousand of these concerts across the nation in towns large and small. Audiences were often in the hundreds, and sometimes in the thousands. Tickets were snatched up and fought over. For many Americans, these were the first, and often the last, professional music performances they ever attended.

Thus, right from the outset, public fascination with the strange new relationship between human bodies and machines created by audio recording rivaled and even exceeded public interest in music itself.

From *How to Record Living Musicians in a Studio?* to *How to Present a Record on Stage?*

The earliest recording devices did not use electrical power. Indeed, record sales had already hit thirty million per year when 90 percent of American homes still had no electricity.[19] Sound waves moving through air were channeled through a large funnel and into a delicate membrane which vibrated in response. A stylus attached to the membrane carved a record of those vibrations into a rotating wax cylinder or platter. To play back the sound, the entire process reversed. The stylus would ride through the grooves carved into cylinder or platter, vibrating the membrane to which it was attached, which created sound waves which were amplified on their way back out the funnel. The rotation of the disc or cylinder was powered by a hand-cranked spring.

The grooves in the disc were cut during the recording process. Once the master was cut, it could not be edited. No mistakes could be corrected, nothing could be spliced out or in. Musicians would gather around the horn of the phonograph and play all together, just as they did in performance. To make one instrument louder than another in a recording, that musician would simply move closer to the horn, or the others would move away.

Electricity changed all of this, first with the introduction of electric microphones and amplifiers in 1925. For the first time, singers did not have to sing loudly to be heard over the background noise of earlier recordings. The "crooner" appeared: men who sang softly into a microphone and let the electronic amplification take care of making their voices heard. Their tender recorded voices tickled listeners' ears with an intimacy that was perceived as almost sexual and made Frank Sinatra the prototype of what would soon become known as a pop star.

Many microphones could be used during a single recording, with the outputs summed electronically in a "mixer." Instead of an ensemble gathering around a horn, each musician could now have a mic placed close to their instrument. Relatively quiet instruments like guitars and string basses could take a leading role, sounding louder on record than even trumpets or trombones. For the first time, the sound that was heard when a record was played back was not anything that could have been heard in the recording studio, or anywhere on earth for that matter. There was nowhere your ear could have been during the making of the record that would have heard what came from the record when played back.

Next came sound recorded onto magnetic tape. Sound was recorded to tape not by cutting grooves in wax but by magnetizing metallic coating on plastic tape in the patterns that mimicked the sound waves. Unlike discs, tape could be cut, and pieces from different recording session could be spliced together. The best bits of multiple takes could be spliced into a whole recording that was better than any single take. Musicians could "play" better on a record than they could play on a stage.

Next came multitrack tape recorders. With a multitrack tape recorder, instruments could be recorded one at a time, yet play back as if they had all been played together. The ensemble never even needed to be present in the studio at the same time. And if the guitar player made a mistake, only the guitar part would be re-recorded, while the other tracks remained unaffected.

Before electricity, musical groups would rehearse on their own, then go into a recording studio. Then the task at hand would be to make a recording that sounded as close as possible to the way the group sounded in a concert. After microphones, amplifiers, tape recording, and multitracking, most records were assembled in a recording studio splice by splice and track by track and only coalesced into what would be heard on the record during the final assembly process. Once the record was released, the musicians would confront the problem of how to take music that they had made fragment by fragment in a studio and present it at a concert.

Instead of a record being compromised version of a live performance, live performance became a compromised version of a recording.

One of the first musicians to grasp the importance of this transition was Canadian piano virtuoso Glenn Gould. Gould was the premier Bach interpreter in the world and one of the most celebrated musicians of the twentieth century, yet in 1964 he announced he would no longer perform concerts. The developments we just discussed made a big impression on Gould. Splicing in the recording studio had become the norm, even for piano virtuosos playing Bach. Gould's recordings, like everyone else's, were made in numerous takes. Both he and the engineer would then listen back to the takes and make a log of which measures were played best in which take. The best sections would then be spliced together to make a recorded "performance" that was technically superior to any single take. Gould was acutely aware that his technical execution in concert could not compete with his spliced together execution on the recorded version of the same piece. He had no wish to go onstage and fail in comparison with his more perfect recorded self, so no more concerts.

Two years later, the Beatles announced that they too would no longer perform concerts. Thus, in the middle of the 1960s, the world's top classical act and top pop act both announced that they would leave the stage for good and disappear into recording studios.[20]

The Snake Swallows Its Tail

The disappearance of concert musicians into the recording studio was mirrored by the appearance onstage of audio play-back devices used as musical instruments.

At the center of these developments were the original DJs, the distant precursors of the guys who get paid hundreds of thousands of dollars to knob sync in funny helmets before hundreds of thousands of fans today.

"DJ" originally referred to "disc jockey," the person who would literally jockey one record after another onto the turntables in the studios of early AM radio stations. Some became celebrity personalities in their own right due to their clever on-air banter. They organized "sock hops" where they would set up a record player on stage and play the same records they were playing on air while teenagers danced (often these events were in gymnasiums, and the dancers wore socks so as not to scuff the floor). Some DJs eventually got their own teen-dance TV shows. Meanwhile in Jamaica, "DJ" came to refer to men who built their own sound systems to blast music to parties they organized. By the 1960s, competing Jamaican DJs were producing records with a production run of just one copy for their own exclusive use at their own events.

In the 1970s, hip hop exploded out of the Bronx, NY, where Grandmaster Flash and others pioneered innovative techniques for hand-manipulating records on turntables. It was here that two turntables on either side of a mixer connected to headphones became the standard DJ setup. The rig facilitated "beatmatching" the tempo and beat of the song currently playing to that of the song next up, eliminating any pause in the music during which people might exit the dance floor. These "scratch" techniques required considerable manual dexterity and practice. By the 1990s, turntable masters like the Invisibl Skratch Piklz had developed these techniques to a stunning degree of virtuosity. But by then the meaning of DJ was already changing yet again, and DJs who stuck with turntables and put in the hours of practice required to scratch fluently were said to be practicing "turntablism." But there was no such thing as turntablism in the 1970s. That was just what hip hop DJs did.

In 1978, Technics released the SL-1200MK2 turntable, specifically designed to facilitate scratching, and Technics turntables have remained the gold standard for dedicated scratchers down to the present day. During those same years, a veritable flood of synthesizers and assorted electronic music instruments were introduced, hyped, and then essentially abandoned. It is curious that the only successful electronic musical instrument we have in the twenty-first century—successful in that it inspires young people to put in the hundreds and thousands of hours to attain a virtuosic level of hand-imagination-ear coordination—is a repurposed nineteenth-century device.

But as those other devices became digital, most DJs abandoned their turntables (if they ever had them) and embraced the computer revolution, creating a global market for "DJ gear" expected to reach nearly $600 million by the end of 2022.[21] At the "Entry-Level DJ" end of that market are devices with a few colored knobs and buttons, along with two larger knobs that are intended to be suggestive of the long-gone turntable. In the midmarket range would be devices with a few more colored knobs and buttons, along with two larger knobs that are intended to be suggestive of the long-gone turntable. And at the high end of the consumer market would be devices with even more colored knobs and buttons, along with the same two larger knobs intended to be suggestive of the long-gone turntable.

The knobs and buttons all serve for data entry into a computer (either integrated into the device or in a laptop set alongside) running software which plays back prerecorded music tracks and alters them in ways which mimic the way Bronx DJs manipulated records by hand back in the 1970s. Want to beatmatch from one track to another? Hit a button. An algorithm will scan the two tracks, detect their tempo, sync them, and align the beats. There are additional buttons for every technique used in the Bronx. There is also an algorithm that determines the musical key of the track that is playing, and displays tracks in your library color coded by musical key, so you can visually scan through options to segue in the same key, or modulate to new ones. Having trouble finding the next track you want to play? An algorithm will search through your entire song library and make suggestions for you, based on its own criteria or criteria you supply. Nothing in your library of thousands of tracks seems right? You can stream from Spotify right into your DJ rig. So pretty much anything that was ever recorded anywhere at any time is grist for your mill.

What if you are more of an EDM-style DJ, and you are more interested in mixing electronic beats than Earth, Wind & Fire tracks? Load up with readymade synthesized sounds. Thousands of these, prelooped for your dance floor pleasure, came bundled with your DJ gear. There are far too many for you to actually listen to in one lifetime, so enter some metadata as search criteria, or let an algorithm make suggestions for you. Want more? Subscribe to one of the DJ streaming services that has every possible kind of looped drums, synthesizers, vocals, whatever—prepackaged at whatever level of finish you want: finished songs; tracks broken out by instrument or sound; or the loops from which those tracks were constructed.

Want to dazzle everyone on the dance floor by playing that rare recording you discovered that only you have? Sorry. There is no rare music anymore. If a recording exists, it is online. And if it is online, it is not rare. Everyone can find it just as easily as you. That's just a fact.

So how do you stand out from the crowd?

A good light show will help. At the upper end of the DJ world, "light show" doesn't really do justice to what's involved. Let's check back in on our friend in the mouse helmet, Joel Zimmerman, aka Deadmau5. In 2018, he released a record of old work, arranged for orchestra by someone else. In 2019, he released a record of old work remixed by other DJs. He recycled old music for two years because he was busy building his Cube, a giant steel-and-carbon-fiber, motorized, video-and-LED-integrated heir to the humble DJ booth. He describes it like this:

> We modeled it in 3D first and then we took it to an engineer and safety engineer and all that. it went through a rigorous and brutal engineering process. . . . The tilt is actually controlled by two hydraulic pistons that expand and contract down there. What happens is that when they pull, they're pulling the whole weight of the cube from the bottom so that's a lot of power. There's a motor down in there, and that junk basically attaches to this monstrosity of a slew ring. So what it does is it holds the ballast all the way back this way. . . . We're getting a lot of bang for our buck in terms of, like, what you would expect to see *if you're going to pay a ticket to go see some fucking guy hit the spacebar.*

The Meaning of Improvisation

Deadmau5 has led us right back to where we were before our digression through recorded music: why would we care that there is a DJ at all? Why isn't a giant steel-and-carbon-fiber cube spewing an overload of sound and light enough? Why do we need the guy in the mouse helmet inside the cube pressing play? Why can't he just press play from the back of the room? Why can't he just press play while wearing his pajamas in bed with a mimosa and send the [play] command in over the internet?

Deadmau5 posed the question, inadvertently, in the interview referenced above when he said, "the best thing of all is the show is 100 percent virtualized so I can take this laptop go home, boot up and develop the whole thing with complete rotating cubes and everything." If everything in the show can be run from a laptop, then the laptop could be inside the cube, in a coffee shop across the street, or on a beach in Bali. All he has to do is type on the laptop keyboard and move his fingers across its trackpad to drive the whole show, from the synthetic bass drum kicks to the tilting of the cube to the mega-every-thing lights and video. Those gestures can be sent to the stage from anywhere there is internet.

Those gestures can be recorded as well. So Zimmerman can sit in a rehearsal sound stage with his entire stage rig fired up and rehearse the show from his laptop, recording every movement of his fingers on the keyboard and trackpad. He can then go back and edit the gestures in bed: perhaps at fifteen minutes and twenty-three seconds the motion on the trackpad should be just a little bit faster; or the keystroke at five minutes and two seconds should have come a hair sooner. He can come back to that sequence, day after day, fine tuning, testing alternative versions, tweaking, until he has everything just how he likes. (A process formerly known as "composing.")

So, why *shouldn't* Zimmerman, or any other DJ, just compose it all exactly how he wants it and then press play? Why does he need to be there, inside the motorized cube? Why doesn't he just (e)mail it in? Stay in bed with a mimosa and the birds chirping outside the window and the sexy babe that all the million-dollar DJs seem to have hanging around in droves. Why press play at the show, where you will blow out your ears with the volume and everything and everyone is annoying, and some kid is going to run up from behind you and startle you in your mouse helmet? Why isn't doing it at the show for suckers?

Because he is *improvising.*

Or at least that is the claim. DJs say they are "reading the room," sensing the energy, making decisions in the collective moment of sound, lights, drugs, dancing, and sex. For the authentic collective EDM experience, these decisions cannot be made in advance and played back. The DJ is the shaman for this most high-tech of rituals. And, yes, shamans must be present and live. They cannot just phone it in.

One might be excused for wondering how connected to an audience of thousands a guy can be if he is hearing not what the thousands of others in the ritual are hearing, but what is coming through the speakers inserted into his ears, which are inserted into the bucket over his head, which is inside a steel-and-carbon-fiber cube, from where he can only see out via a chest-mounted video camera.

But the idea that the DJ is improvising, sensing the crowd and responding accordingly, is challenged not only by the exalted circumstances of the million-dollar DJs. The trope is challenged even more forcefully by the circumstances of humblest of DJs, the ones who have become ubiquitous all over the world in countries rich and poor; north, south, east, and west; in shopping malls and hotel lobbies and corner bars.

I have seen them at men's clothing stores promoting sales events. And even though the shoppers tried their best to ignore them, there the DJs were: diligently at their task; one hand holding a headphone to one ear; the other hand on a knob; face locked in an expression of utter concentration. Busy busy. Man at work. Stand back.

Or in my neighborhood gay bar in San Francisco on a Wednesday night. Which a few years back might have been full of guys feeling festive and on the make, but now all the hot guys are back at home swiping through profiles on smartphone hookup apps, so the only people at the bar are a few lonely old men crying into their beer. Yet there is the DJ, with the same headphones, the same knobs, the same expression. Sensing the room. Making decisions in the moment.

(If I were DJing that room and trying to read the energy of the crowd, I would be tempted to play taps. Oddly enough, taps is a piece of music that will figure prominently later in this essay.)

Can't he make an iTunes playlist during his coffee break earlier in the day, and then just hit play at the bar? Why would he even need to make a playlist? Spotify must have a whole series of playlists appropriate for cheering up lonely men at empty bars on Wednesday nights. There are probably searchable meta tags for that. But, no, he cannot do any of that because he has to make those decisions live, in the moment. In other words, he has to *improvise*.

Improvisation, Repetition, and Loops

"Improvisation" was a hot topic in Western music from the mid-twentieth century until recently, though its story played out differently depending on the context.

In Europe, the evolution of musical practice has been deeply intertwined with the evolution of the system for notating it with symbols on a page. The earliest notation merely indicated whether the pitch of a note should be higher or lower than the previous note, rather than a particular pitch, and there was no indication at all regarding rhythm. Rather than a complete set of musical instructions, early musical scores were more like sets of reminders to assist with performances everyone mostly knew from memory. Over the centuries, scores became more and more detailed as the role of the composer became more and more exalted, and the musical resources of instrumental ensembles became more and more complex. Eventually the composer had become a great man, who retreated alone to a great tower, to think his great thoughts and report them back to the not-great folk, by way of a score which precisely specified each instruction they must follow to realize his great idea. By the beginning of the twentieth century, the score itself had become the perfect realization of the composition, and performances were heroic attempts to match that perfection but ultimately doomed to fall short. By the second half of the twentieth century, this had all become such a musical jail that young composers trying to make their mark were turning to the idea of improvisation with renewed interest.

Black American music took a very different course.[22] From its roots in Africa and then through centuries of slavery, black music was passed down from one generation to the next orally. Young musicians learned by watching and listening to others, not by looking at symbols on paper. In the same sense that there was no "live music" before there was recorded music, improvisation was not a big issue in black American music because there was no big idea of compositions fixed in every detail.

And then, at the start of the twentieth century, along came audio recording technology. If you want to argue that technologies have no meaning in and of themselves but only acquire it from the social context in which they land, audio recording technology is your Exhibit A, because its consequences for white and black music were quite opposite.

Oral musical traditions evolve slowly. Imagine an itinerant blues guitarist traveling around the American South before the advent of records. If you want to learn to play like him, you go see him every chance you get, which is not very often. Remember, there are no recordings to listen to, so unless you are in his presence, you don't get to hear him play. If you live in an out-of-the-way place, you may never hear him at all, but only hear of him.

Now add records. You get a record by that same guitarist and it's like he is visiting your town every day and every hour, whenever you want him. You can play along with the record on your guitar, playing the same song, or even the same phrase, over and over, days on end, until you have it mastered. Every little innovation anyone introduces can be quickly generalized across the entire community, community in this case meaning all those who have access to the same records. How did this musician do this? Or that? Let's listen again. Someone comes up with a new variation, records it, and the cycle repeats.

This opened a flood of musical innovation in black American music, from the beginning of recording until about the 1960s, which has few equals in history. From Jelly Roll Morton to Louis Armstrong to Charlie Parker to John Coltrane to Ornette Coleman in just a few decades. From New Orleans to swing to bebop to modal jazz to free jazz. Breathtaking. And absolutely impossible without recording.

Recording had the opposite effect in the white musical tradition that traced its roots back to Europe, and in which the score was predominant. Before recording, there was a certain freedom to "interpret" the classic scores. Performances were fleeting, existing only until the concert ended. The transitory nature of the performance left room to experiment and try novel approaches. There was no recording to which every other performance of the same work could be compared. The permanent, more important realization of a composition was the score. The appearance of records led to a race to create the definitive recording of this or that great piece of music. Once a definitive recording was agreed upon, individual performances were expected to measure up. Only musicians of exceptional confidence could buck this system.

Then along came tape splicing and the jail door closed even tighter. Now you could record multiple takes of a piece, comb through them measure by measure, and splice all the best bits into a whole that made for a definitive recorded version which no musician could actually play. Not even the musician who had played on the recording. This is the point at which Glenn Gould decided to retire from the stage.

But now it's the 1960s. You have a new kind of person, the "teenager," listening to a new kind of music, "pop" or "rock." The whole notion of the teenager as a kind of person, who is largely defined by the music they listen to, which is not the music their parents listen to—all of this was made possible by music technology. Portable music technology, to be specific. Before the transistor radio and the portable record player, the radio and record player were great big things in the living room. The whole family had to listen to the same records or broadcasts, just as the whole family heard the same pieces played on the piano before record player and radio displaced the piano. It wasn't until recording technology became sufficiently inexpensive and portable that the kids could go off on their own and make their own choices about what to listen to that music became a defining part of adolescence. So you get Frank Sinatra, and then Elvis and Little Richard, and then the Beatles, and by the time you get to the 1960s you get Jimi Hendrix playing before the Woodstock nation.

The 1960s were permeated by the idea of freedom. Freedom from segregation, freedom from the military draft, free love, free expression, free music. Vietnam fighting for freedom from imperialism, students in Prague demanding freedom from Communism, students in Paris, Mexico City, and Berkeley demanding freedom from capitalism. Feminists demanding freedom from patriarchy. Gays at the Stonewall riots demanding freedom from homophobia. And so on.

It was in this fertile terrain that musicians of the white European tradition seeking freedom from the score, musicians from the black tradition playing free jazz, and young people playing extended "jams" in psychedelic rock bands, all crossed paths in pursuit of freedom through musical improvisation. Miles Davis played extended electronic jams in the same vein as Jimi Hendrix, while Karlheinz Stockhausen (Germany's latest "great composer") released improvised records using filters and distortion devices similar to those used in acid rock.

This was the musical world I came of age in. The idea of musical improvisation was everywhere. It was much more than a musical style. It was an idea around which you might organize your entire life. Maybe it would free your mind. Maybe it could change the world. Somehow, the idea and practice of musical improvisation seemed subversive to power structures much bigger than those of the musical world. This may sound like a hippie idea, but it was shared by John Cage and Sun Ra. A hopelessly naive idea in retrospect, but compelling at the time.

I entered that world playing a synthesizer. Not the kind manufactured today for DJs, but an early modular synthesizer. The kind that looked vaguely like a telephone switchboard, with wires all over the place. No keyboard, just a lot of knobs and buttons. Mostly those things lived in the research labs of large universities, not in concert halls or clubs. I was not the first person to take one on stage and play it as a performance instrument, but I was the first person to make the core of my musical practice using a synthesizer like that to play "free improvisation" with others playing guitars, saxophones, drums, and so on. I did so night after night in the downtown New York City music scene of the late 1970s, with John Zorn, Eugene Chadbourne, Vernon Reid, Leslie Dalaba, Toshinori Kondo, and others.

It made sense to use a synthesizer like that in an improvisational setting, because you never knew exactly what sound was going to come out of those things. If you made something good one day, you tried to record it right away because you would never be able to recreate it exactly. The controls were not sufficiently precise. Improvisation was a technical necessity.

Being the only person improvising on stage in the 1970s with an electronic instrument that had no keyboard but just knobs and buttons, surrounded by instrumentalists, was an awkward place to be. Inevitably, someone would make the criticism that I was *not really playing* an instrument. Often the person voicing the criticism would be another musician. And it was true that, whatever I was doing, it was different from what the musicians around me were doing. I had no special virtuosity at turning knobs that resulted from hours upon months upon years of practice. I couldn't turn a knob better than anyone else. If I was good at playing my synthesizer, it was because I had a deep understanding of how it worked technically, and I had spent hundreds of hours listening closely to the details of its sounds, and I understood how to pick out the most interesting patterns of sound and use the knobs to manipulate all of that in musically interesting if not precisely predictable ways. But there was nothing particularly interesting about watching me turn the knobs, in the way that it is inherently interesting to watch a master violinist play a violin. I was acutely aware of all this, and whenever possible I would go on stage with other musicians who were unequivocally *playing* their instruments.

Today's DJs, with their DJ gear that has lots of knobs and buttons but no keyboard, are in a sense my present-day heirs. They are closer to that precedent than they are to the turntable DJs of the 1970s, who developed such nuanced physical dexterity in the manipulation of records. DJ gear, however, is quite opposite of the 1970s synthesizers in one aspect: the ease with which the exact same sounds can be repeated, loop after loop, hour by hour, night after night. In fact, it is next to impossible to coax that gear into doing anything *but* precise repetition. Press a button and start a loop, or a looped series of loops, or a looped series of a looped series of loops. Each one exactly the same, each time through the loop. Same same same same same. Press another button and capture those perfectly repeating loops inside of perfectly repeating loops, and loop that. Press another button and record the whole set of loops. Press another button and you can post the recording on the web the moment your set is over. Now everyone with access to the web can download and loop it. And then loop that loop.

Throughout history, music has had repeating elements. A song might have two verses and two choruses. A round is a song form created from repeating musical phrases. Without any sense of repetition in time, you cannot have a sense of musical rhythm. But *repetition*, whether it involves Afro-Cuban percussionists playing off a repeating clave, or Bach interpreters following the repetition makings in the score, is not the same as an electronic device playing a *loop*. There will always be nuanced variation in musical repetition as executed by human bodies. We simply cannot repeat motions absolutely precisely with our muscles, no matter how much we practice. And anyway, why would anyone want to? The nuanced variation in musical repetition is what makes human music feel human. Electronic devices are just the opposite. Superhumanly precise repetition is the easiest option. Adding nuanced variation is more complex. Hence, the loop.

The idea of improvisation, as we understood it in the 1970s, has lost most of its juice. Few musicians today feel the need to break out of the constraints imposed by musical scores. In fact, fewer and fewer musicians even learn how to read musical notation. I have given guest lectures at music schools where most of the students in attendance cannot read music and have no interest in learning. And it is no longer radical to propose that you could "compose" music by going into a recording studio without a lot of worked out ideas, messing around with the technology, and seeing what happens, then splicing the best bits together. It is how most records are made. Billie Eilish and her older brother did that in their bedroom and swept the 2020 Grammy Awards with the results.

And the meaning of improvisation?

Improvisation is now the last claim of the living human being to be present on the concert stage. Why is the DJ up there among all those automatic loops run by machine? Because he is supplying the human element: improvisation, meaning deciding which loops run for how long and at what volume. That, the DJ would have you believe, is the one piece of the process a machine cannot do as well as a human.

Surprisingly, modular synthesizers like I used to play in the 1970s have meanwhile made a comeback. There are many many more of them in the world today, and more people who use them, than there ever were back in the day before they were replaced by digital synthesizers. What makes this even more remarkable is their cost. A hardware modular synthesizer that goes for $10,000 can have a software equivalent that runs on a smartphone and costs $30. Yet young musicians are shelling out their hard-earned cash for the expensive hardware and then lugging them around to gigs at dance clubs. They speak of how they want to touch real knobs and wires, not move their fingers across laptop trackpads while watching images of knobs and wires on a screen.

I understand wanting to touch a real guitar string or hit a real drumhead. You touch the "real" thing that actually vibrates and makes sound. But what is "real" about turning a knob? The knob doesn't make sound. It is connected to a circuit, which is powered by electricity, which is sent down a wire to a speaker box, in which a piece of cardboard vibrates to create the "real" sound. No one touches the cardboard. The idea that touching a knob is more real than a trackpad is a telling measure of how mediated by electronics our basic sense of music has become.

The hardware synthesizers do have an enormous advantage over their software competitors: you cannot get your email on them. Or check Facebook with them. You can't take selfies with them. Or watch porn on them. You can't call anyone with them. You don't have to remember a password to open them. No one can hack into them. And they will not constantly ask you to download their latest upgrade.

The laptops and smartphones that run software synthesizers have become devices that produce more distraction than anything else. Young people who sit down at a modular synthesizer find that they can concentrate on music in a way that is new for them. Apparently, this is worth thousands of dollars.

Lizzo's Flute and the Army's Bugle

Acoustic musical instruments are not going away. People still learn to play them. In fact, it may be that people are playing them better than ever. Several friends who teach at top music schools have commented to me that they had never before seen students coming through their schools with the skill sets of today. They attribute this in large part to how much music young people are exposed to in the age of YouTube and Spotify. Young musicians come up listening to everything that has ever been recorded: Indian ragas, Indonesian gamelan, Afropop, Rachmaninoff, Tuvan throat singing, bebop masters, Swiss yodelers, Nine Inch Nails, and Yo-Yo Ma. They hear it all. Their curiosity is limited only by the hours in the day, not, as in the case of my own youth, by the few records that were available at the only record store in a small Colorado farming town.

But music involving instrumental virtuosity is becoming a niche item, like handmade shoes. In most of the world and for hundreds of years, people have worn shoes. And since shoes were made by hand, shoemakers were everywhere. All of that changed with the application of mass industrial manufacturing techniques to shoes. Sure, there are still some people who really, really like handmade shoes and are willing to pay extra for them. And there are still practicing shoemakers who cater to that niche market of handmade shoe aficionados. But for most of us, factory-made shoes are just fine. And much cheaper. And if you use machines,[23] you can do some things in the making of shoes that are just not possible by hand.

Musical instrumentalists are going the way of the shoe-makers. Yes, there is still an audience that is willing to pay extra to watch someone actually blow into a trombone or scrape a bow across the strings of a contrabass. And, as with shoemakers, there are still people who are drawn to put in the thousands of hours required to master these instruments. But most people don't want to go to all that trouble when, without any practice at all, they can press a button and hear a similar sound. Or many sounds. Before electronics, there were a limited number of ways to create pitched sound in the lower register of human hearing loud enough to be useful in a musical ensemble. Tubas are hard to beat in this regard. Their mechanisms for generating loud pitched bass notes without electricity are ingenious. But they are extremely awkward to play. Now you can press a button on an electronic device and call forth a great variety of pitched bass notes, at any volume you like, from much softer than you could play on a tuba to much louder. Convincing young people to learn to play a tuba is, understandably, a harder and harder sell.

Meanwhile, acoustic instruments are treated with reverence in those rare moments when they do appear. Recently I saw a concert by the irrepressible rapper and singer Lizzo. She had a cohort of dancers with her who were very much present in flesh and blood, but no musicians, just a recorded backing track over which she sang and rapped. The fact that she did not engage in any between-song banter until late in the show suggested that someone backstage had hit play at the outset and the backing track had run uninterrupted through most of the show. All of Lizzo's PR materials claim that, in addition to singing and rapping, she is a "classically trained flutist." And late in the set, one of her dancers exited stage right and returned bringing forth the mysterious instrument as if it was a holy relic. The dancers literally bowed down. Lizzo took the flute and played a few rudimentary lines. The crowd went nuts. After a few measures, the flute was gingerly carried away and the recorded backing track continued unmolested. On the way out of the theater after the concert, I heard numerous amazed fans exclaiming, "And she can even play the flute!"

The US armed forces can train its personnel to service a drone, fire a Multi-role Anti-armor Anti-personnel Weapon System, or disarm an improvised explosive device, but training its personnel to play taps on a bugle has become a reach too far. So military funerals now feature a uniformed bugler who, at the appointed moment, solemnly brings a bugle to the lips and presses a button which triggers a recording of a bugle playing taps, which plays on a speaker stuffed into the bugle's horn.

The US military started playing taps at funerals during the Civil War. About 140,000 Union troops died in combat in that war. Another 53,000 in World War I and 300,000 in World War II. But there were always enough bugle players.

The bugle has no valves like a trumpet or slide like a trombone. Consequently, the bugle is only capable of producing five notes. It is one of the simplest musical instruments in the world.

The digital bugle manual, in its entirety, reads as follows:

- Switch the unit on with the on/off switch. The red light will illuminate.
- The volume control knob can be set from normal to extra loud.
- When you press the play button, the green LED will illuminate, this warns you that taps will start in 5 seconds and the bugle should be placed to the mouth.
- Once taps has finished playing the green light will go out. The bugle should then be turned off by the on/off switch.
- Battery Warning Light: If the red LED starts to flash the batteries must be replaced.[24]

What Is Live Music?

So, what do we mean when we say "live music?" I have been asking this question of musicians I meet around the world.

> "If I can bring things [loops] in and out when I want to, it's live. Like, if I can take the vocals out, or bring them back in. Or take the kick drum out, or bring it back in, then I am playing live."
> —Club DJ

OK, I picked that quote pretty much at random, but variations on that are the most common answer I have received.

> "If a performance can fail on aesthetic rather than technical grounds, it is live."
> —Chris Cutler, drummer, essayist, and record distributor

Nice one. Very thoughtful.

> "If you are staring at a screen, you are not playing live."
> —Geo Wyeth, musician and performance artist

I love this answer. It was completely unexpected but made immediate sense. We all have had this experience: we are conversing with someone who pulls out a phone and tries to nonchalantly text using one part of their brain while distractedly conversing with us using another. Our perception of these moments is that once someone is looking at the screen, they have absented themselves from the conversation. They are no longer "live."

After listening to people answer this question for several years, I have come to think that any answer that references who or what is making the music is rooted in the past. For people coming of age today "live music" refers not to how the music is made but to how it is experienced. If the music is experienced among a crowd of other living people, the music is live.

To grasp how compelling live music in this sense has become, think of what has happened to movies. Movie theaters are empty. Everyone is at home streaming movies online. Some people who came up during the heyday of movie theaters bemoan the loss of the collective experience of appreciating movies in a crowd. But the collective aspect of watching movies in theaters was always muted. Yes, we did leave our homes, go to the theater, stand in line, and then sit down among a crowd. But then the lights went out and, for the most part, watching the movie was a solitary experience. The exception was comedy, because even in the dark we could hear each other laugh. Comedies suffer from the watching-at-home experience. It is rare that we laugh out loud when we are alone, even when watching a very funny movie. Maybe we chuckle a time or two. Put that same comedy in a movie theater and laughter ensues. In fact, the amount of laughter is proportionate to the density of the audience. The same movie that gets a chuckle or two at home, will get some laughs in a sparsely attended theater, and roaring laughter in a packed theater.

Many people now have technology for watching movies at home which compares favorably with a theater: large, high resolution screens, surround sound speakers, the whole bit. And also a cozy sofa, a fully stocked kitchen in the next room, and a pause button for bathroom breaks. By contrast, no one has a sound system at home that can remotely compare with the sound systems at dance clubs and festivals.

The collective experience of dance clubs and festivals involves more than high decibels and heavy bass. First and foremost is the experience of dancing itself, which remains a deeply social activity. Few people dance at home alone. Does anyone anywhere in the world put on a Deadmau5 track their apartment and dance alone? Also included in the dance club package is the collective sweat of hundreds or thousands or even hundreds of thousands of people. And alcohol and MDMA and molly and cocaine and ketamine. And sexual possibility and sometimes sex. And smiling and laughing and knowing that you and the people around you are engaged in a transcendent collective endeavor. And you are flush with the confidence you are in the know, dancing to the sub-micro-micro genre of electronic dance music that is the coolest thing going right now, and even if climate change and coral reef death and the new authoritarianism are all unstoppable, you are sharing an ecstatic moment you have created together, which is in some way a giant collective "fuck you" to all of that.

Try doing that at home alone.

In 2020 that is live music.

If That's All There Is, Then Let's Keep Dancing

In the 1980s, as electronic drum machines became cheaper, smaller, easier to program, and mass manufactured, a joke began circulating among touring musicians:

> Q: How many drummers does it take to screw in a lightbulb?
> A: None, they've got machines that do that now.

No one imagined back then that just a decade or two later, there would be giant "dance music festivals" around the world, lots of them, branded into microgenres of styles, at which hundreds of thousands of people would dance for days and nights on end, and not a single drummer would be present. Dance music festivals are a drummer-free zone.

If the sight of a living actor on a movie set is "an orchid in the land of technology," today's dance music festivals are deserts with no plants at all.

There are, of course, still drummers in those parts of the world dominated by electronic dance music. You can hear them in rock bands, jazz groups, and so on—nearly always in places and contexts in which people don't dance. You want to find people dancing today, go to where the drummers aren't.

Anyone who has given the matter any thought can quickly tell whether a given recording was made with a human drummer or a machine. Machines divide time into precise units, while humans divide time imperfectly. This is what gives human drummers a "human sound," and what makes electronic loops so different. Many people may still prefer to hear human rhythm in rock music and other genres. But if they want to dance, they want a machine keeping time.

This is not limited to the world of electronic dance music. I grew up in ranch country, two-stepping to country music, and every now and then I still like to go country-and-western dancing. There is a large and active C&W dancing club in San Francisco where I live. It may well be that more people go C&W dancing today in San Francisco than ever before—always to *recorded* C&W music.

One time I asked the president of the club why they only had recorded music and never a group of musicians.

"Because you can't dance to it."

"Say what?"

"No, really. We tried it. We've had live bands here. We couldn't dance to it. Everyone complained. To do the line dances and all, the music has to be exactly like it was on the record. The bands just don't get it."

Burning Man

Eighty thousand people gather in a desolate desert in Nevada once a year for a week of dancing and drugs known as Burning Man. There are "sound camps," which are stationary installations of massive sound and light systems, and "art cars," which are buses or trucks outfitted with massive sound and light systems. The advantage of sound camps is that ultimately you can fit more lights and speakers on the ground than you can on the most outlandishly tricked out truck. The advantage of art cars is that they move.

As the sun sets and the desert turns to night, the lights and sound come on and thousands of burners load up on drugs and head out into the desert, moving from sound camp to sound camp, or following an art car as it rolls slowly through the darkness. All the music is loud. Not just loud, but LOUD. Everyone present is engaged in a collective sacrifice of their future ability to hear. At some level they all know that, and the collective nihilism is part of the tribal bonding. If their boss subjected them to noise like that loud at work they could sue his ass. And they would. But out here in the desert, they *choose* to sacrifice their hearing. Fuck the future. This is live.

The music at the different camps and trucks can sound nearly identical to the uninitiated. but the burners know it is differentiated by BPM, or beats per minute, a technical expression for tempo that comes from music software and leaked into the public dance music lexicon. A decade or four ago, music lovers discussed which music had more swing, or more funk, or rocked harder. All those terms refer to highly nuanced ways that human bodies making music divide time into rhythm. Machines divide time into precisely equal units, which vary only in speed. The music can go faster or slower. Thus, BPM. And burners *know* their BPM. This camp runs the BPM at 120. If that is not your thing, the camp over there runs at 116. Completely different feel.

At Burning Man 2019, all the music was "bass music," the latest sub-micro-micro genre of electronic dance music that features one or two bass notes repeated over and over and over and over and over. Bass music is a wonderful example of technology's unexpected consequences. Beginning with the rock festivals of the 1960s, young people gathered in large numbers to listen to loud music, creating a demand for sound systems that could accurately reproduce sound at high volume across a large space. One of the technical hurdles involved how to accurately produce bass frequencies at extreme volume. Above a certain volume, the bass loses definition and gets mushy. So a lot of R&D went into how to build massive sound systems with a tight, punchy low end, with impressive results.

Once the new systems existed, people started messing around with them and discovered that if you pump high volume bass frequencies around forty hertz (the lowest octave of a piano) through a system like that, they resonate strongly with a human body. If the tones start abruptly, they will literally punch you in the gut. Not just in the gut, you will feel it all over. "Bass music" is all about that punch. Dancing becomes something different. Your whole body gets slammed at whatever BPM the DJ selected in the software. Your body cannot help but react. How you react to those full body slams becomes your particular way of dancing.

You don't do this sort of dancing with a human partner. Everyone has the same partner: the machine. To put it in terms of partner dancing like tango or salsa or waltzing, the machine is the "lead" dance partner. All the humans are "following." The machine does what it does, everyone else reacts. Follow-dancing with a machine, out on the desert, for long hours, with tens of thousands of others, and a collection of drugs in your body, is a deeply meaningful experience for many people.

The visual focal point of every sound camp and art car is a sort of altar from which the DJ commands the system. The visual reference to a church altar is often explicit. DJs take turns at these altars. The music barely changes from one DJ to the next, though the core members of the tribe, the ones most deeply knowledgeable of its ways, do take note. And Deadmau5 was right when he said that "given about 1 hour of instruction, anyone with general knowledge of music tech" can do it. Think of it like driving a sports car that fits thousands of people. It's a hot car, accelerates in no time, turns on a dime. Fun. Anyone who knows how to drive can drive it, make it go faster, slower, and turn, so any rider who wishes to do so can take a turn at the wheel. It's fun. If everyone is on drugs, it's a blast.

Soul Entrainment

The notion of entrainment provides an interesting way to think about dancing. Broadly speaking, "entrainment" refers to how two independent rhythmic processes can interact in such a way as to eventually synchronize. The process was first noted in 1666 by the inventor of the pendulum clock, who noticed that pendulums mounted on the same board eventually synchronize. The idea has been applied to a number of fields, from chemistry to geology. Biological entrainment ranges from how living organisms synchronize to the rising and setting of the sun (synchronizing to a rhythm they have no influence over) to how fireflies flash together (mutual synchronization).

Humans have the nearly unique ability to mutually synchronize to an external isochronous pulse (a pulse produced outside their bodies at a regular interval of time). And we can do more than that: we can *infer* a regular beat from a more complex pattern of sound which does not include the actual beat among its audible elements. Indeed, we can hardly stop ourselves. Give us a nice groove, and we will start tapping our toes, in sync with each other, regardless of what languages we speak, how prominent a role dancing plays in our cultures, and so on. Give us just a little encouragement and we will soon be dancing. Like language, this ability is generalized across our species. No matter how high or low we score on whatever sort of intelligence test, no matter how educated or unschooled we are, nearly all of us can learn language and learn to dance.[25]

We have been dancing for as long as we have been human, and we have used this ability to collectively entrain to a beat to great effect. We dance to learn who our tribe is, who our equals and superiors and inferiors are, how to overcome fear, how to share in joy and sorrow, and how to love. From slaves in fields to armies at battle to sailors heaving rope on ships, we use a beat as a tool to aggregate effort.

What, if anything, is lost or gained when we stop entraining to each other and entrain to machines instead? Maybe nothing. Perhaps to suggest otherwise is to succumb to a romanticism about the "nature" of human bodies that is based only in nostalgia. But it is difficult to stand in the dirt at Burning Man, in the midst of thousands of people entrained to the same machine, surrounded on every side by different groups of thousands of people entrained to their slightly different machines, each crowd of thousands facing a wall of lights and speakers with an altar reserved for the person who actually gets to touch the machine, and believe that nothing profound has changed from when we entrained to each other, to beats created by other human bodies. What's more, it is hard not to connect this change with the many other ways our attention is shifting away from each other and toward technology, from our inability to tear our attention away from our phones, to our inability to turn away from bizarrely authoritarian political leaders who rule by tweet. When people coming of age entrained their bodies to other human bodies, they were learning to live in a world in which relations with other humans was the central life experience. If young people today come of age by entraining to machines, perhaps they are learning how to live in a world in which their relations with machines will be paramount.

What Time Is It?

In 1931, Lewis Mumford published one of the first major histories of technology, *Technics and Civilization*.[26] Mumford uses the word *technics* (from the ancient Greek word *tekhne)* to refer to not only what we call technology but also the skill, knowledge, and social organization required to create and operate it; in sum, the interplay between culture, society, and technology. Mumford argued that "the clock, not the steam-engine, is the key machine of the modern industrial age." His argument cuts to the heart of the matters we are considering here, so it is worth considering his thinking in detail:

> [The clock] gave the human enterprise the regular collective beat and rhythm of the machine; for the clock is not merely a means of keeping track of the hours, but of synchronizing the actions of men.... In its relationship

to determinate quantities of energy, to standardiza-
tion, to automatic action, and finally to its own special
product, accurate timing, the clock has been the foremost machine in modern technics.... It disassociated
time from human events and helped create the belief in
an independent world of mathematically measurable
sequences....

There is relatively little foundation for this belief
in common human experience: throughout the year
the days are of uneven duration, and not merely does
the relation between day and night steadily change, but
a slight journey from east to west alters astronomical
time by a certain number of minutes. In terms of the
human organism itself, mechanical time is even more
foreign. While human life has regularities of its own,
the beat of the pulse, the breathing of the lungs, these
change from hour to hour, and in the longer span of
days, time is measured not by the calendar but by the
events that occupy it. The shepherd measures from the
time the ewes lambed; the farmer measures back to the
day of sewing or forward to the harvest....

To keep time was once a peculiar aspect of music:
it gave industrial value to the workshop song or the
chanty of the sailors tugging at a rope. But the effect
of the mechanical clock is more pervasive and strict: it
presides over the day from the hour of rising to the hour
of rest. When one thinks of the day as an abstract span
of time, one does not go to bed with the chickens on a
winter's night: one invents wicks, chimneys, lamps, gas-
lights, electric lamps, so as to use all the hours belonging
to the day. When one thinks of time, not as a sequence
of experiences, but as a collection of hours, minutes,
and seconds, the habits of adding time and saving time
come into existence. Time took on the character of an

enclosed space: it could be divided, it could be filled up, it could even be expanded by the invention of labor saving instruments.

Abstract time became the new medium of existence. Organic functions themselves were regulated by it: one ate, not upon feeling hungry, but when prompted by the clock: one slept, not when one was tired, but when the clock sanctioned it.[27]

Following Mumford, we could say that dancing to beats measured in electronically precise units of time is nothing new, it is just the latest step in a centuries-long transition from experiencing life in organic or biological time, to mechanical or automated time.

Computers are, of course, clocks. Very precise clocks. The latest iteration of the clocks of which Mumford wrote. Clocks in 1934 divided time into hours, minutes, and seconds. The chip in the current iPhone divides one second into 2.65 billion equal units.

One way to think about it would be to say that a musical beat was a good way to aggregate human labor back when clocks could not divide time into units finer than those humans could tap their feet to (BPMs). Now that we can divide a second in billions of equal units, rates at which a human body cannot keep pace, aggregating tiny electrical signals that can actually work at that speed becomes more important than aggregating the work of human muscles.

Finally, if it is true that the ability to collectively entrain to isochronous beats (as opposed to, say, fireflies, which can mutually entrain to each other) is uniquely human, then it should come as no surprise that we choose to dance to the most rigorously isochronous beats available. In fact, if this ability to collectively entrain to isochronous beats, whether a marching song or a sailor's chanty or the punch clock at the factory, forms the foundation upon which we have built this high-tech society, then mechanical time, followed by digital time, is the most human of inventions.

Glenn Gould's Participant Listener

Technological change always involves elements that are perceived as a loss by some and a gain by others. Older people whose lives have been entangled with the world as they found it are more likely to see mostly loss, while younger people whose attention is focused on possibility gravitate toward the gain.

I am drawn to thinkers who call attention to precisely what is changing and what is staying the same, and to how that change might be steered in a way that benefits the many and harms the few. I have special respect for those who called attention to historic changes long before their full scope was apparent. Even better if they managed to find democratic potential in those changes. And if they embraced that potential despite its threat to their own position of privilege, they are my heroes.

I have very few such heroes. Consider Karl Marx. Marx anticipated the vast social political changes that industrial production would bring in its wake, and he did so when there were very few factories in the world. He foresaw enormous negative consequences but also how the massive disruptions might benefit the many. But as extraordinary as his accomplishment was (and how ultimately tragic its shortcomings), none of this threatened his own privilege.

With regard to the changes in music technologies discussed here, I do have a hero, but my hero is not one of the twentieth century's celebrated musical radicals. My hero is Glenn Gould, the piano virtuoso who stopped performing in 1966. Gould wrote an extraordinary essay explaining that decision, which did not get much attention either at the time or in the years since.[28] In the world of "classical music," the essay was dismissed as the ramblings of an eccentric genius. Among the "avant-garde," no one even noticed. The latest thinking of the world's greatest Bach interpreter was not on their reading list.

In very much the same way that Marx looked at the nascent factory system and extrapolated out, Gould thought deeply about the recording studio. We have already noted his conclusion that the tape splice permitted recordings to boast a degree of technical perfection unattainable in a concert performance, even by the musician who had made the recording, and how this led him to stop performing. But the great pianist followed the implications of the tape splice far beyond its immediate impact on his own musical practice, and well into the future.

> [Decision-making] privileges will not need to remain the exclusive preserve of a tape editor. . . . They could quite possibly be delegated directly to the listener. It would indeed be foolhardy to dismiss out of hand the idea that the listener can ultimately become his own composer. . . .
>
> Forty years ago the listener had the option of flicking a switch inscribed "on" and "off" and, with an up-to-date machine, perhaps modulating the volume just a bit. Today [1966], the variety of controls made available to him requires analytical judgment. And these controls are but primitive, regulatory devices compared to those participational possibilities which the listener will enjoy once current laboratory techniques have been appropriated by home playback devices.[29]

Gould makes an extraordinary intellectual leap here concerning the coming movement of laboratory techniques into the home. (Of course now they have moved from our homes to our pockets). He did so decades before computer-based audio, sampling keyboards, and web streaming. Robert Moog and Don Buchla had only recently put the finishing touches on their very first synthesizers when Gould wrote those words.

> At the center of the technological debate, then, is a new kind of listener—a listener more participant in the

musical experience. The emergence of this mid-twentieth century-phenomenon is the greatest achievement
of the record industry. For this listener is no longer pas-
sively analytical; he is an associate whose tastes, preferences, and inclinations even now alter peripherally the
experiences to which he gives his attention, and upon
whose fuller participation the future of the art of music
waits.[30]

From today's vantage point a half century later, we no longer
have to squint to see Gould's participant listeners, nor await
their fuller participation in the art of music. They are everywhere, with their headphones, playlists, and knobs. We call
them DJs.

It would be a relatively simple matter, for instance, to
grant the listener tape-edit options which he could exer-
cise at his discretion. Indeed, a significant step in this
direction might well result from that process by which
it is now possible to disassociate the ratio of speed to
pitch and in so doing . . . splice segments of [music]
recorded at different tempos. . . . This process could, in
theory, be applied without restriction to the reconstruc-
tion of musical performance. There is, in fact, nothing
to prevent a dedicated connoisseur from acting as his
own tape editor and, with these devices, exercising such
interpretive predilections as will permit him to create
his own ideal performance.[31]

It took a decade after Gould made this prediction for Grand
Master Flash working in The Bronx to figure out how to
seamlessly splice music recorded at different tempos. It took
another decade for samplers to appear that could do the same
digitally. And it was only in the last few years that digital audio
software became powerful enough to disassociate the ratio of
speed to pitch.

Today's DJs do all of this with the push of a button. Algorithms embedded in their gear calculate the tempo and key of different songs and shift them to match. Gould anticipated all of this in 1966 while watching technicians cut recording tape with razor blades and splice the pieces together with scotch tape: "All the music that has ever been can now become a background against which the impulse to make listener-supplied connections is the new foreground."[32] Had Gould been writing today instead of half a century ago, he could not have been more precise in depicting the situation of the twenty-first century DJ at his gear with its built-in web streaming, headphones in one hand, the other hand scrolling through long lists of tracks sorted by metadata, foregrounding his listener-supplied connections against a background of all music that has ever been.

But Gould has more. He considers not only how technology will change music and how it is made, but also how it will change the hierarchies of power in which those who make, sell, and listen to music are embedded. For Gould's "participant listener":

> is also, of course, a threat, a potential usurper of power, an uninvited guest at the banquet of the arts, one whose presence threatens the familiar hierarchical setting of the music establishment. Is it not, then, inopportune to venture that this participant public could emerge untutored from that servile posture with which it paid homage to the status structure of the concert world and, overnight, assume decision-making capacities which were specialists' concerns heretofore?[33]

Who is threatened here? Whose power is this uninvited guest at the banquet of the arts poised to usurp? Glenn Gould's, that's whose. And all other virtuosos whose gift of acute musical hearing is intertwined with an equally extraordinary physical dexterity of which the rest of us can only dream. Gould, however, is delighted by the prospect that the hierarchy he sits atop might topple, because he foresees a more democratic, participatory future taking its place.

> The most hopeful thing about this process . . . [is that] this whole question of individuality in the creative situation . . . will be subjected to a radical reconsideration . . . If these changes are profound enough, we may eventually be compelled to redefine the terminology with which we express our thoughts about art. Indeed, it may become increasingly inappropriate to apply . . . the word "art" itself. In the best of all possible worlds, art would be unnecessary. Its offer of restorative, placative therapy would be begging for a patient. The professional specialization involved in its making would be presumption. The generalities of its applicability would be an affront. The audience would be the artist and their life would be art.

There was perhaps no one in the world who stood to lose more from these developments than Glenn Gould. His position is finally far more radical than those we often think of as the luminaries of the twentieth century avant-garde. Even John Cage, the great disrupter of musical modernism, never relinquished his position as great composer, despite his insistence that sounds thrown together by chance were as interesting as sounds put together by choice.

Like revolutionaries everywhere, Gould maintained a stubborn optimism regarding the changes he observed. A half century on, we are in a position to ask whether his optimism was warranted.

I suspect that many of the developments about which I have expressed skepticism would have been welcomed by Gould. Take, for example, the assertion by Deadmau5 that, "Given about 1 hour of instruction, anyone . . . could DO what im doing at a deadmau5 concert."

Isn't that the more participatory world Gould looked forward to? What's wrong with that? When musicians harbor doubts about all of this, is that merely evidence that we are clinging to our privilege more tightly than Gould did?

Yes, technology has put thousands of instrumentalists out of work, replaced by legions of DJs, but would Gould mind?

"As the medium evolves," Gould hoped, "[the] quite properly self-indulgent participation of the listener will be encouraged," while the "class structure within the musical hierarchy . . . will become outmoded." He looked forward to the collapse of "our post-Renaissance orgy of musical sophistication" and a return to something like "the medieval status of the musician [as] one who created and performed for the sake of his own enjoyment."[34]

However, Gould completely failed to anticipate that the act of listeners supplying their own connections between their favorite recordings would become a performance activity, much less a platform for superstars surrounded by personality cults.

I suspect Gould would have been delighted with the bottom-of-the-food-chain DJs at clothing stores, shopping malls, and corner bars, listeners who participate in the music-making process far more than the passive listeners of his day. But I do wonder what he would have had to say about the million-dollar DJs, faces hidden in helmets, pumping their fists in the air before pressing play and then knob syncing in front of tens of thousands of adoring fans who paid extravagant fees to attend.

The Negative Space at Her Side

If the question is whether, in the midst of all the technologies of sound and light and we need a DJ at all, then Japanese pop sensation Hatsune Miku has an emphatic answer: we don't.

Miku began existence as a "Vocaloid," or software voice synthesizer. People who buy the Hatsune Miku software can enter their own melodies and lyrics on their computers, and the software generates a synthetic voice singing their song. There is no vast library under the hood of recordings of someone singing every possible word or syllable in the Hatsune Miku repertoire. The software generates the voice, which sounds very much like a computer. No one is fooled into thinking the voice is human.

All Hatsune Miku had at her birth was a voice algorithm, a pigtailed anime image with a lot of leg on the box cover, an age (she was sixteen the day she was born), and a tempo and pitch range she preferred (70–150 BPM, with a vocal range from A below Middle C to an E two octaves up). Miku's creators released no songs or videos, just the software. Then she exploded.

As of early 2020, she had released over 100,000 songs, making her by far the most prolific pop star in history. All the songs were created by Hatsune Miku aficionados and posted on social media. None of this seems to have been directed, or even fully anticipated, by the company that sells the software. Miku also has, at last count, 170,000 YouTube videos and over a million images online. These are the result of a 3D art and animation program created by an independent programmer and given away for free. She has over two million fans on Facebook. She has a real-world race car team, a licensing deal with computer game giant Sega, and massive "expos" around the world full of fan-designed merch.

In the early days of computer art, many of us who were experimenting with the early tools wondered if someday this this technology would allow for artistic collaborations with a previously unimaginable number of participants. That day has arrived.

Surely Glenn Gould would be pleased. Hatsune Miku's legions of fans are indeed "a new kind of listener—a listener more participant in the musical experience . . . whose presence threatens the familiar hierarchical setting of the music establishment. . . . The whole question of individuality in the creative situation is subjected to a radical reconsideration. . . . The audience is the artist."[35]

Then she started performing as a 3D hologram at sold-out concert venues around the world. Tokyo. Hong Kong. New York. Paris. London. Amsterdam. Jakarta. Mexico City. She has been a soloist with major symphony orchestras. Devoted fans show up by the thousands, often coming hours early to check out the merch and bask in the glow. The literal glow. They all bring official Hatsune Miku glow-sticks which they wave in coordinated ways, while some wait to see if she will sing the song they wrote or throw down a dance move they choreographed.

In 2014, she opened for Lady Gaga. On either side of her hologramatic presence, a flesh-and-blood girl danced in choreographed hypergeneric J-pop unison behind a screen. Hyper bright generic pop concert videos were projected on the screens from behind. Dancing between the projectors and the screens, the dancers' bodies broke the light beams and were thus visible mostly as shadows. Living bodies as negative space, a black hole in a blaze of light

Cell phone videos from the show quickly made it to YouTube, where the comments section bristled with complaints from Miku fans who took offense at the behavior of Lady Gaga's audience:[36]

> It makes me sad that the audience doesn't even look like they are enjoying her

> The lack of crazed dancing and fainting is a crime against humanity. A crime against Miku. A crime against art.

> i genuinely wanna know what was going on in the audiences mind lol

> the American audience really bugs me. nobodys into it at all. i would of been screaming and crying and would

of bought tickets just to see MIKU. AND I WOULD OF BEEN DANCING AND SINGING ALONG LIKE HELL YEAH.

Everyone was so confused omg. The least they could've done was clap along when she asked them too,

poor miku baby, these people have no idea wat they are seeing. everyone there must be like wtf is this, but *so many people would die to see her live.*

Hatsune Miku live.

Acknowledgments

There are so many people I have to thank for their contributions to this book.

First of all, the long-time members of what I think of as my intellectual and emotional executive committee: Pierre Hébert, James Magee, Christian Huygen, Marlena Sonn, and Ryan Nguyen. And others who gave invaluable feedback and criticism regarding the essays in this book: Marian Ekweogwu, Chris Cutler, Fred Frith, Joshua Clayton, Seth Horvitz, Yan Jun, Etienne Noreau-Hebert, and Geo Wyeth. And in particular Jon Leidecker. My extended back-and-forth conversations with Jon were fundamental in the development of the final chapter.

Facebooking the Anthropocene in Raja Ampat recounts my experiences during a fifteen-month world concert tour. Musicians I played with on the tour include Theo Bleckmann, Audrey Chen, Fred Frith, Gerry Hemingway, Mazen Kerbaj, Phil Minton, Henrik Munkeby Nørstebø, Jon Rose, Sharif Sehnaoui, and Rully Shabara.

The tour could not have happened without the support of many people who went far out of their way to make it happen. At the top of the list is overall tour manager Marc Kate. Josh Feola coordinated the Chinese leg of the tour as well as three different concerts in Beijing, and hosted me in Beijing along with Emma Xiaoming Sun. Ariele Monti coordinated the Italian leg of the tour and, along with his beautiful collaborators at Area

Sismica, hosted a concert in Forli. Hosts and concert promoters in specific cities include: Sacco V. in Shanghai; Auttaratt Photongnoppakun in Bangkok; Siew-wai Kok and Yandsen in Kuala Lumpur; Nicolaus Mesterharm in Phnom Penn; Gabby Miller of the Queer Forever Festival in Hanoi; Tengal in Manila; Martinus Indra Hermawan and the Jogja Noise Bombing collective in Yogyakarta; Mbuh Adin and the Kolektif Hysteria in Semerang; Donny Hendrawan in Malang; Jonas Sestakresna in Denpasar; the Ruangrupa collective in Jakarta; Anitha Silvia in Surabaya; Reza Enem and Mirwan Adnan in Makassar; Myrna Melgar Sr., Marisol Galindo, and Myrna Melgar Jr. in El Salvador; Walter Schmidt in Mexico City; Miguel Galperin in Buenos Aires; Luis Alvarado in Lima; Benjamin Vergara in Valparaiso; Nandy Cabrera in Montevideo; Rebecca Mazzei and Joel Peterson in Detroit; Jim Stahley and Gage Boone in New York City; Hans Falb in Nickelsdorf; Cynthia Placha in Prague; Dragan Ambrozic in Belgrade; Pedro Rocha in Porto; Martin aka Snediggen Snurssla in Brno; Mathias Maschat and everyone at Ausland in Berlin; Francesco Giomi and everyone at Tempo Reale in Florence; Luca Venitucci and Fabrizio Spera in Rome; Fabrizio Elvetico, Giuseppe Aiello, Ciro Longobardi and everyone from L'Asilo in Naples; Sharif Sehnaoui in Beirut; and people in cities and towns around the world who hosted concerts but whose names I do not have at hand.

New friends I made during the tour and who supported me in these endeavors in one way or the other include Dede Oetomo, Wukir Suryadi, Paulina Rojas, Julian "Togar" Abraham, Yanri Subronto, and Mami Vinolia.

Big thanks to Ramsey Kanaan and Joey Paxman and all their colleagues at PM Press. And to my daughter Katie Miles, and to my partner in life and love, Grace Towers.

Notes

Facebooking the Anthropocene in Raja Ampat

1   Michael Slezak, "Coral Bleaching Event Now Biggest in History—and About to Get Worse," *Guardian*, June 20, 2016, https://www.theguardian.com/environment/2016/jun/21/coral-bleaching-event-now-biggest-in-history-and-about-to-get-worse.

2   Joby Warrick, "The Persian Gulf Could Be Too Hot for Human Survival by 2090. Here's What This Means for Your City," *Washington Post*, October 28, 2015, https://www.washingtonpost.com/news/energy-environment/wp/2015/10/26/climate-change-could-soon-push-persian-gulf-temperatures-to-lethal-extremes-report-warns/.

3   Associated Press, "Vietnam Blames Toxic Waste Water from Steel Plant for Mass Fish Deaths," *Guardian*, July 1, 2016, https://www.theguardian.com/environment/2016/jul/01/vietnam-blames-toxic-waste-water-fom-steel-plant-for-mass-fish-deaths.

4   J.A. Zalasiewicz, "The Geological Cycle of Plastics and Their Use as a Stratigraphic Indicator of the Anthropocene," *Anthropocene* 13 (2016): 4–17.

5   Ana Swanson, "How China Used More Cement in 3 Years than the U.S. Did in the Entire 20th Century," *Washington Post*, March 24, 2015.

6   Christina Gough, "Value of the Global Video Games Market from 2012 to 2021," *Statista*, September 14, 2018.

7   Libby Hogan in Yangon and Michael Safi, "Revealed: Facebook Hate Speech Exploded in Myanmar during Rohingya Crisis," *Guardian*, April 2, 2018, https://www.theguardian.com/world/2018/apr/03/revealed-facebook-hate-speech-exploded-in-myanmar-during-rohingya-crisis.

8   Agence France-Presse, "Indian State Cuts Internet after Lynchings over Online Rumours," *Guardian*, June 29, 2018, https://www.theguardian.com/world/2018/jun/29/indian-state-cuts-internet-after-lynchings-over-online-rumours; Hannah Ellis-Petersen, "Social Media Shut

Down in Sri Lanka in Bid to Stem Misinformation," *Guardian*, April 21, 2019, https://www.theguardian.com/world/2019/apr/21/social-media-shut-down-in-sri-lanka-in-bid-to-stem-misinformation.

9 Dylan Byers and Ben Collins, "Trump Hosted Zuckerberg for Undisclosed Dinner at the White House in October," *NBC News*, November 20, 2019.

10 Paul Mozur, "In Hong Kong Protests, Faces Become Weapons," *New York Times*, July 26, 2019.

11 Bob Moore (@BobMooreNews), "When you overlay @CBP stats with county population figures, it's startling," Twitter, March 21, 2019, 12:27 p.m., https://twitter.com/BobMooreNews/status/1108812466060959746.

12 Mitch Smith, Julie Bosman, and Monica Davey, "Flint's Water Crisis Started 5 Years Ago. It's Not Over," *New York Times*, April 25, 2019.

13 Merritt Kennedy, "2 Former Flint Emergency Managers, 2 Others Face Felony Charges over Water Crisis," *NPR*, December 20, 2016, https://www.npr.org/sections/thetwo-way/2016/12/20/506314203/2-former-flint-emergency-managers-face-felony-charges-over-water-crisis.

14 Erica L. Green, "Flint's Children Suffer in Class after Years of Drinking the Lead-Poisoned Water," *New York Times*, November 6, 2019.

15 IQAir, AirVisual, http://airvisual.com, accessed December 26, 2019.

16 Linda Poon, "Why Indonesia Wants to Move Its Capital out of Jakarta," *CityLab*, May 6, 2019.

17 Stanley Widianto, "Crumbling Seawall Heightens Worries over Flood Threat to Indonesian Capital," *Reuters*, December 13, 2019, https://br.reuters.com/article/us-indonesia-seawall-idUSKBN1YH14U.

18 Tim van Emmerik, "Riverine Plastic Emission from Jakarta into the Ocean," *IOP Science*, August 13, 2019.

19 Steve Allen, Deonie Allen, Vernon R. Phoenix, et al., "Atmospheric Transport and Deposition of Microplastics in a Remote Mountain Catchment," *Natural Geoscience* 12, no. 5 (2019): 339–44; Melanie Bergmann et al., "White and Wonderful? Microplastics Prevail in Snow from the Alps to the Arctic," *Science Advances* 5, no. 8 (August 14, 2019).

20 Paul Rogers, "Levels of Plastic Pollution in Monterey Bay Rival Those in Great Pacific Garbage Patch," *Mercury News*, June 6, 2019, https://www.mercurynews.com/2019/06/06/monterey-bay-is-littered-with-tiny-pieces-of-plastic-groundbreaking-new-study-finds/.

21 Stephen Buranyi, "The Missing 99%: Why Can't We Find the Vast Majority of Ocean Plastic?," *Guardian*, December 31, 2019, https://www.theguardian.com/us-news/2019/dec/31/ocean-plastic-we-cant-see.

22 Terry P. Hughes et al., "Global Warming Impairs Stock-Recruitment Dynamics of Corals," *Nature*, April 3, 2019.

23 Gemma Conroy, "'Ecological Grief' Grips Scientists Witnessing Great Barrier Reef's Decline," *Nature*, September 13, 2019.

24 Karen B. Roberts, "Coral Reef Starter Kit," *University of Delaware Daily*, October 14, 2019.

Are Two Dimensions Enough?

1 James Vincent, "Former Facebook Exec Says Social Media Is Ripping Apart Society," *The Verge*, December 11, 2017.

2 David Karf, "25 Years of WIRED Predictions: Why the Future Never Arrives," *Wired*, September 18, 2018.

3 Heather Hopkins, "Telemedicine in Arizona," *Institute for Metal Health Research*, 2016.

4 Kirk Johnson, "TV Screen, Not Couch, Is Required for This Session," *New York Times*, June 8, 2006.

5 Larry Hendricks, "Take Two Pills and Tune in Tomorrow," *Arizona Daily Sun*, June 2, 2006.

6 Johnson, "TV Screen."

7 Todd Kendall, "Pornography, Rape, and the Internet," Department of Economics, Clemson University, 2006.

8 Robyn Greenspan, "Porn Pages Reach 260 Million," *Internetnews.com*, September 25, 2003, https://www.internetnews.com/bus-news/article.php/3083001/Porn+Pages+Reach+260+Million.htm.

9 Kendall, "Pornography, Rape, and the Internet."

10 World-Wide-Mind project website, www.w2mind.org/.

11 J. Cheetham, "The Perils of Porn," *Sydney Morning Herald*, February 3, 2005.

12 Mireya Navarro, "The Most Private of Makeovers," *New York Times*, November 2, 2004.

13 Author's interview, January 30, 2007. The interviews were conducted in person in San Francisco and online with men around the US. All interviews were conducted on the condition of anonymity.—Ed.

14 Author's interview, February 3, 2007.

15 Regina Lynn, *The Sexual Revolution 2.0: Getting Connected, Upgrading Your Sex Life, and Finding True Love—or at Least a Dinner Date—in the Internet Age*, Berkeley: Ulysses, 2005.

16 "Who knew the time you spent jacking off would be minimized by the time you spend categorizing your damn porn?" Author's interview, February 3, 2007.

17 "Futurologists" is not my word but theirs. Try a Google search on it.

18 Author's interview, February 3, 2007.

19 Here are some of the postings from San Francisco's Craigslist (sfbay.craigslist.com) on one random day, January 2, 2007:

- "Looking for a guy that has a nice cock I can worship while you watch some hot porn."
- "Got porn here playing (got lots of it) . . . Just looking for other hotties to share with, watch with and play with."
- "Cum over, watch hot porn, let me suck your cock."
- "Just got back in and looking for buds to come on by, strip down, and watch some porn."
- "CIRCLE JERK FOR COOL DUDES WATCHIN STR8 PORN!"
- "JACK OFF TO PORN AT YOUR PLACE."
- "Lookin for a bud to kick it with. watch some porn, stroke, suck and see what else is fun for two horny guys."
- "Looking to watch porno and mutual J/O only."
- "Wanna watch porno on my laptop . . . I bring my laptop, and we watch some mindblowing scenes I have accumulated over time. These scenes are guaranteed to show a thing or two you might never have thought of doing. Then if our chemistry is right, we can take it from there."

20   Author's interview, December 2, 2007.
21   https://www.secondlife.com, accessed December 30, 2006.
22   Wallace, Mark, "A Virtual Holiday in the Virtual Sun," *New York Times*, October 28, 2005.
23   Christine Lagorio, "Is Virtual Life Better Than Reality?," *CBS Evening News*, July 31, 2006.
24   Seth Schiesel, "They Got (Video) Game; N.B.A. Finals Can Wait," *New York Times*, June 21, 2005.
25   Schiesel, "They Got (Video) Game."
26   Schiesel.
27   Sherry Turkle, "Whither Psychoanalysis in a Computer Culture?," *Psychoanalytic Psychology* 21, no. 1 (October 23, 2004): 16–30.
28   Furby Connect, http://furby.hasbro.com/.
29   Ylan Q. Mui, "Teddy Bear, Version 2.0," *Washington Post*, February 7, 2007.
30   Mui, 2007.
31   See Stereo3D, http://www.stereo3d.com/hmd.htm#chart.
32   Andrew Wallenstein, "Dialogue: Media Tunes in Second Life," *Hollywood Reporter*, February 13, 2007.
33   All details of the case of Roy Brown from Fernanda Santos, "With DNA from Exhumed Body, Man Finally Wins Freedom," *New York Times*, January 24, 2007.
34   Joseph Weizenbaum. *Computer Power and Human Reason: From Judgment to Calculation*. San Francisco: W.H. Freeman, 1976.
35   Kenneth Mark Colby, James B. Watt, and John P. Gilbert, "A Computer Method of Psychotherapy: Preliminary Communication," *Journal of Nervous and Mental Disease* 142, no. 2 (1966): 148–52.

36 Noah Wardrip-Fruin and Nick Montfort, *The New Media Reader* (Cambridge, MA: MIT Press, 2003), 370.

37 SimilarWeb, https://www.similarweb.com/top-websites.

38 "The 2019 Year in Review," *Pornhub Insights*, December 11, 2019.

39 George Szalai, "Video Game Industry Growth Still Strong: Study," *Reuters*, June 21, 2007.

40 Adam Green, "Global Video Games Market Expected to Reach $138 Billion by 2021," *Golden Casino News*, November 26, 2019.

41 Ryan Schultz, "Second Life Infographic: Some Statistics from 15 Years of SL," https://ryanschultz.com/2018/04/23/second-life-infographic-some-statistics-from-15-years/.

42 "Video Gamers & Pornhub," *Pornhub Insights*, December 9, 2013.

43 Jane Coaston, "The Proud Boys, the Bizarre Far-Right Street Fighters behind Violence in New York, Explained," October 15, 2018, https://www.vox.com/2018/10/15/17978358/proud-boys-gavin-mcinnes-manhattan-gop-violence.

44 Jacob Bernstein, "How OnlyFans Changed Sex Work Forever," *New York Times*, February 19, 2019. This section also draws on informal discussions between the author and several "fans-only content creators."

45 Simon Hattenstone, "The Rise of eSports: Are Addiction and Corruption the Price of Its Success?," *Guardian*, June 16, 2017, https://www.theguardian.com/sport/2017/jun/16/top-addiction-young-people-gaming-esports.

46 Jack Morse, "How a Dead Veteran Became the Face of a Therapy App's Instagram Ad," *Mashable*, December 19, 2019.

47 Talkspace, https://www.talkspace.com, accessed December 27, 2019.

48 "Privacy Policy," Woebot Health, https://www.woebot.io/privacy, accessed December 27, 2019.

49 Morse, "How a Dead Veteran Became the Face of a Therapy App's Instagram Ad."

Technics Turntables and Civilization

1 Albert Watson, "The Rise of Deadmau5," *Rolling Stone*, June 20, 2012.

2 "Deadmau5 Admits 'We All Hit Play,'" *BrooklynVegan*, June 25, 2012.

3 Ashley Cullins, "Deadmau5 Sued over 'Meowingtons' Trademark," *Hollywood Reporter*, March. 14, 2017.

4 Ashley Zlatopolsky, "Deadmau5 on Why EDM Is Finished, Learning to Love His New Album," *Rolling Stone*, December 7, 2016.

5 Indra Pajaro, "Inside the Fairy Tale Festival: How Does Tomorrowland Work?," *Medium*, July 21, 2018; for the four hundred thousand figure, see https://en.wikipedia.org/wiki/Tomorrowland_(festival)#2019.

6 Ellie Mullins, "The Secrets That Tomorrowland Usually Hide," We Rave You, July 10, 2019, https://weraveyou.com/2019/07/tml-secrets/.

7   Marina Muhlfriedel, "Electrifying the EDM Environment with Lighting and Stage Designer Steve Lieberman," 2017, https://pro.harman.com/insights/entertainment/touring/electrifying-the-edm-environment-with-lighting-and-stage-designer-steve-lieberman/.

8   Check *DJ Magazine*'s list of the top one hundred DJs: https://djmag.com/top100djs, or *Forbes*'s top fifteen DJs: https://www.forbes.com/sites/monicamercuri/2019/07/29/the-worlds-highest-paid-djs-of-2019/#2c6874f27a97.

9   The early outliers were pipe organs and music boxes. In pipe organs, the air forced through the pipes would come from a bellows operated by hand. Not the hand of the person playing the organ keyboard, but of a dedicated bellows operator. In music boxes, the power came from a coiled spring cranked by hand.

10   "The Piano's History, Quietly Celebrated," *New York Times*, November 24, 1988.

11   "Tin Pan Alley," *Acoustic Music*, https://acousticmusic.org/research/history/musical-styles-and-venues-in-america/tin-pan-alley/, accessed November 20, 2019.

12   Gerald Carson, "The Piano in the Parlor," *American Heritage* 17, no. 1 (December 1965): 54–59.

13   *The Dinah Shore Chevy Show*, January 20, 1963.

14   "'After the Ball': Lyrics from the Biggest Hit of the 1890s," History Matters, http://historymatters.gmu.edu/d/5761/.

15   "U.S. Sales Database," Recording Industry Association of America, https://www.riaa.com/u-s-sales-database/.

16   "Historic American Sheet Music," Digital Scriptorium, Duke University, https://library.duke.edu/rubenstein/scriptorium/sheetmusic/browse.html, accessed November 23, 2019.

17   "History of the Record Industry, 1920–1950s Part Two: Independent Labels, Radio, and the Battle of the Speeds," *Medium*, June 8, 2014, https://medium.com/@Vinylmint/history-of-the-record-industry-1920-1950s-6d491d7cb606, accessed November 20, 2019.

18   Advertisement for Bush Music House and the Edison Corporation, *Penn Yan Democrat*, October 27, 1922.

19   "History of the Record Industry, 1877–1920s Part One: From Invention to Industry," *Medium*, https://medium.com/@Vinylmint/history-of-the-record-industry-1877-1920s-48deacb4c4c3, accessed November 20, 2019.

20   Gould and the Beatles did not quit performing for identical reasons. For the Beatles, the primary reason, in addition to security concerns, seems to have been that the concert sound systems of the day could not compete with the screaming fans at Beatles concerts, so the group literally could no longer hear themselves play. Nevertheless,

it remains striking how the top classical and pop acts quit performing at the same time, and both disappeared into recording studios where they pioneered many studio techniques that remain in use to the present day.

21 "Global DJ Equipment Market 2019 by Manufacturers, Regions, Type and Application, Forecast to 2024," 360 Reports, February 6, 2019, https://www.360researchreports.com/global-dj-equipment-market-13836795.

22 Words and terminology are a thorny problem in discussions like this. Boundaries are always difficult to place precisely, and there were white people playing "black music" and black people playing "white music." But the US is, to this day, a largely segregated society, and its music has been largely segregated as well. I could use terms like "classical music" and "jazz," but those come with their own problems. I know black musicians who passionately argue that "jazz" is an idea invented by white people for the purpose of marginalizing black music.

23 By "machines" I mean any automated device powered by something other than human muscle that does work. Music machines, by this definition, include any device that runs on electric power that produces automate sound.

24 Ceremonial Bugle, http://www.alabamavva.org/bugle.html.

25 There are individual animals of other species that have demonstrated this ability, but thus far research has not confirmed another species in which this ability is generalized. The Thai Elephant Orchestra, however, requires a careful examination.

26 Lewis Mumford, *Technics and Civilization* (New York: Harcourt, Brace, 1934).

27 Mumford, *Technics and Civilization*, 14–17.

28 Glenn Gould, "The Prospects of Recording," *The Glenn Gould Reader* (New York: Vintage, 1990), 331–52.

29 Gould, "The Prospects of Recording," 347.

30 Gould, 347.

31 Gould, 348.

32 Gould, 350.

33 Gould, 347.

34 Gould, 351–52.

35 Gould, 347.

36 Hatsune Miku Full Opening for Lady Gaga, May 20, 2014, St. Paul, MN, uploaded May 22, 2014, YouTube, https://www.youtube.com/watch?v=OYlRN6XWsDE&t=1321s, accessed December 23, 2019.

Index

"Passim" (literally "scattered") indicates intermittent discussion of a topic over a cluster of pages.

About the Author

Bob Ostertag's work cannot easily be summarized or pigeon-holed. He has published more than twenty albums of music, five books, and a feature film. His writings on contemporary politics have been published on every continent and in many languages, beginning with his work as a journalist covering the civil war in El Salvador in the 1980s. His books cover a wide range of topics, from labor unions to the history of journalism to estrogen and testosterone. He has performed at music, film, and multimedia festivals around the globe. His radically diverse musical collaborators include the Kronos Quartet, John Zorn, Mike Patton, transgender cabaret icon Justin Vivian Bond, British guitar innovator Fred Frith, EDM DJ Rrose, and many others.

ABOUT PM PRESS

PM Press is an independent, radical publisher
of books and media to educate, entertain, and
inspire. Founded in 2007 by a small group of
people with decades of publishing, media, and
organizing experience, PM Press amplifies the
voices of radical authors, artists, and activists.
Our aim is to deliver bold political ideas and vital stories to all walks
of life and arm the dreamers to demand the impossible. We have sold
millions of copies of our books, most often one at a time, face to face.
We're old enough to know what we're doing and young enough to know
what's at stake. Join us to create a better world.

PM Press
PO Box 23912
Oakland, CA 94623
www.pmpress.org

PM Press in Europe
europe@pmpress.org
www.pmpress.org.uk

FRIENDS OF PM PRESS

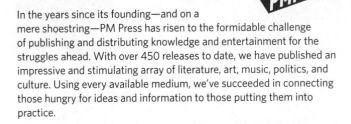

These are indisputably momentous times—the financial system is melting down globally and the Empire is stumbling. Now more than ever there is a vital need for radical ideas.

In the years since its founding—and on a mere shoestring—PM Press has risen to the formidable challenge of publishing and distributing knowledge and entertainment for the struggles ahead. With over 450 releases to date, we have published an impressive and stimulating array of literature, art, music, politics, and culture. Using every available medium, we've succeeded in connecting those hungry for ideas and information to those putting them into practice.

Friends of PM allows you to directly help impact, amplify, and revitalize the discourse and actions of radical writers, filmmakers, and artists. It provides us with a stable foundation from which we can build upon our early successes and provides a much-needed subsidy for the materials that can't necessarily pay their own way. You can help make that happen—and receive every new title automatically delivered to your door once a month—by joining as a Friend of PM Press. And, we'll throw in a free T-shirt when you sign up.

Here are your options:

- **$30 a month** Get all books and pamphlets plus 50% discount on all webstore purchases

- **$40 a month** Get all PM Press releases (including CDs and DVDs) plus 50% discount on all webstore purchases

- **$100 a month** Superstar—Everything plus PM merchandise, free downloads, and 50% discount on all webstore purchases

For those who can't afford $30 or more a month, we have **Sustainer Rates** at $15, $10 and $5. Sustainers get a free PM Press T-shirt and a 50% discount on all purchases from our website.

Your Visa or Mastercard will be billed once a month, until you tell us to stop. Or until our efforts succeed in bringing the revolution around. Or the financial meltdown of Capital makes plastic redundant. Whichever comes first.

Anthropocene or Capitalocene? Nature, History, and the Crisis of Capitalism

Edited by Jason W. Moore

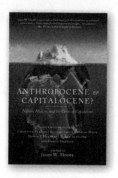

ISBN: 978-1-62963-148-6
$21.95   304 pages

The Earth has reached a tipping point.
Runaway climate change, the sixth great extinction of planetary life, the acidification of the oceans—all point toward an era of unprecedented turbulence in humanity's relationship within the web of life. But just what is that relationship, and how do we make sense of this extraordinary transition?

Anthropocene or Capitalocene? offers answers to these questions from a dynamic group of leading critical scholars. They challenge the theory and history offered by the most significant environmental concept of our times: the Anthropocene. But are we living in the Anthropocene, literally the "Age of Man"? Is a different response more compelling, and better suited to the strange—and often terrifying—times in which we live? The contributors to this book diagnose the problems of Anthropocene thinking and propose an alternative: the global crises of the twenty-first century are rooted in the Capitalocene; not the Age of Man but the Age of Capital.

Anthropocene or Capitalocene? offers a series of provocative essays on nature and power, humanity, and capitalism. Including both well-established voices and younger scholars, the book challenges the conventional practice of dividing historical change and contemporary reality into "Nature" and "Society," demonstrating the possibilities offered by a more nuanced and connective view of human environment-making, joined at every step with and within the biosphere. In distinct registers, the authors frame their discussions within a politics of hope that signal the possibilities for transcending capitalism, broadly understood as a "world-ecology" that joins nature, capital, and power as a historically evolving whole.

Contributors include Jason W. Moore, Eileen Crist, Donna J. Haraway, Justin McBrien, Elmar Altvater, Daniel Hartley, and Christian Parenti.

If It Sounds Good, It Is Good: Seeking Subversion, Transcendence, and Solace in America's Music

Richard Manning with a Foreword by Rick Bass

ISBN: 978-1-62963-792-1
Price: $26.95   320 pages

Music is fundamental to human existence, a cultural universal among all humans for all times. It is embedded in our evolution, encoded in our DNA, which is to say, essential to our survival. Academics in a variety of disciplines have considered this idea to devise explanations that Richard Manning, a lifelong journalist, finds hollow, arcane, incomplete, ivory-towered, and just plain wrong. He approaches the question from a wholly different angle, using his own guitar and banjo as instruments of discovery. In the process, he finds himself dancing in celebration of music rough and rowdy.

American roots music is not a product of an elite leisure class, as some academics contend, but of explosive creativity among slaves, hillbillies, field hands, drunks, slackers, and hucksters. Yet these people—poor, working people—built the foundations of jazz, gospel, blues, bluegrass, rock 'n' roll, and country music, an unparalleled burst of invention. This is the counterfactual to the academics' story. This is what tells us music is essential, but by pulling this thread, Manning takes us down a long, strange path, following music to deeper understandings of racism, slavery, inequality, meditation, addiction, the science of our brains, and ultimately to an enticing glimpse of pure religion.

Use this book to follow where his guitar leads. Ultimately it sings the American body, electric.

"Richard Manning is the most significant social critic in the northern Rockies. We're fortunate to have Dick Manning as he continues his demands for fairness while casting light on our future."
—William Kittredge, author of *The Last Best Place: A Montana Anthology* and *The Next Rodeo: New and Selected Essays*

The Explosion of Deferred Dreams: Musical Renaissance and Social Revolution in San Francisco, 1965–1975

Mat Callahan

ISBN: 978-1-62963-231-5
$22.95   352 pages

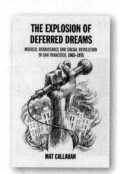

As the fiftieth anniversary of the Summer of
Love floods the media with debates and celebrations of music, political
movements, "flower power," "acid rock," and "hippies", *The Explosion
of Deferred Dreams* offers a critical re-examination of the interwoven
political and musical happenings in San Francisco in the Sixties.
Author, musician, and native San Franciscan Mat Callahan explores
the dynamic links between the Black Panthers and Sly and the Family
Stone, the United Farm Workers and Santana, the Indian Occupation of
Alcatraz and the San Francisco Mime Troupe, and the New Left and the
counterculture.

Callahan's meticulous, impassioned arguments both expose and reframe
the political and social context for the San Francisco Sound and the
vibrant subcultural uprisings with which it is associated. Using dozens
of original interviews, primary sources, and personal experiences,
the author shows how the intense interplay of artistic and political
movements put San Francisco, briefly, in the forefront of a worldwide
revolutionary upsurge.

A must-read for any musician, historian, or person who "was there" (or
longed to have been), *The Explosion of Deferred Dreams* is substantive and
provocative, inviting us to reinvigorate our historical sense-making of an
era that assumes a mythic role in the contemporary American zeitgeist.

*"Mat Callahan was a red diaper baby lucky to be attending a San Francisco
high school during the 'Summer of Love.' He takes a studied approach, but
with the eye of a revolutionary, describing the sociopolitical landscape that
led to the explosion of popular music (rock, jazz, folk, R&B) coupled with
the birth of several diverse radical movements during the golden 1965-1975
age of the Bay Area. Callahan comes at it from every angle imaginable
(black power, anti-Vietnam War, the media, the New Left, feminism, sexual
revolution—with the voice of authority backed up by interviews with those
who lived it."*
—Pat Thomas, author of *Listen, Whitey! The Sights and Sounds of Black
Power 1965-1975*

Silenced by Sound: The Music Meritocracy Myth

Ian Brennan
with a Foreword by Tunde Adebimpe

ISBN: 978-1-62963-703-7
$20.00   256 pages

Popular culture has woven itself into the social
fabric of our lives, penetrating people's homes
and haunting their psyches through images
and earworm hooks. Justice, at most levels, is something the average
citizen may have little influence upon, leaving us feeling helpless and
complacent. But pop music is a neglected arena where concrete change
can occur—by exercising active and thoughtful choices to reject the low-
hanging, omnipresent corporate fruit, we begin to rebalance the world,
one engaged listener at a time.

Silenced by Sound: The Music Meritocracy Myth is a powerful exploration
of the challenges facing art, music, and media in the digital era. With his
fifth book, producer, activist, and author Ian Brennan delves deep into his
personal story to address the inequity of distribution in the arts globally.
Brennan challenges music industry tycoons by skillfully demonstrating
that there are millions of talented people around the world far more
gifted than the superstars for whom billions of dollars are spent to
promote the delusion that they have been blessed with unique genius.

We are invited to accompany the author on his travels, finding and
recording music from some of the world's most marginalized peoples.
In the breathtaking range of this book, our preconceived notions of art
are challenged by musicians from South Sudan to Kosovo, as Brennan
lucidly details his experiences recording music by the Tanzania Albinism
Collective, the Zomba Prison Project, a "witch camp" in Ghana, the
Vietnamese war veterans of Hanoi Masters, the Malawi Mouse Boys,
the Canary Island whistlers, genocide survivors in both Cambodia and
Rwanda, and more.

Silenced by Sound is defined by muscular, terse, and poetic verse, and
a nonlinear format rife with how-to tips and anecdotes. The narrative
is driven and made corporeal via the author's ongoing field-recording
chronicles, his memoir-like reveries, and the striking photographs that
accompany these projects.

After reading it, you'll never hear quite the same again.

Save the Humans?
Common Preservation in Action

Jeremy Brecher

ISBN: 978-1-62963-798-3
$20.00   272 pages

We the people of the world are creating the
conditions for our own self-extermination,
whether through the bang of a nuclear holocaust or the whimper of an
expiring ecosphere. Today our individual self-preservation depends on
common preservation—cooperation to provide for our mutual survival
and well-being.

For half a century Jeremy Brecher has been studying and participating in
social movements that have created new forms of common preservation.
Through entertaining storytelling and personal narrative, *Save the
Humans?* provides a unique and revealing interpretation of how social
movements arise and how they change the world. Brecher traces a path
that leads from the sitdown strikes on the pyramids of ancient Egypt
through America's mass strikes and labor revolts to the struggle against
economic globalization to today's battles against climate change.

Weaving together personal experience, scholarly research, and historical
interpretation, Jeremy Brecher shows how we can construct a "human
survival movement" that could "save the humans." He sums up the
theme of this book: "I have seen common preservation—and it works."
For those seeking an understanding of social movements and an
alternative to denial and despair, there is simply no better place to look
than *Save the Humans?*

"*This is a remarkable book: part personal story, part intellectual history
told in the first person by a skilled writer and assiduous historian, part
passionate but clearly and logically argued plea for pushing the potential
of collective action to preserve the human race. Easy reading and full of
useful and unforgettable stories. . . . A medicine against apathy and political
despair much needed in the U.S. and the world today.*"
—Peter Marcuse, author of *Cities for People, Not for Profit: Critical Urban
Theory*